Floor Style

Floor Style

Yvonne Rees
Technical Adviser: Tony Herbert

 VAN NOSTRAND REINHOLD
New York

A QUARTO BOOK

Copyright © 1989 Quarto Publishing plc

Library of Congress Catalog Card Number 89-5787
ISBN 0-442-23923-8

Published in the U.S.A. by
Van Nostrand Reinhold
115 Fifth Avenue
New York, New York 10003

This book was designed and produced by
Quarto Publishing plc
The Old Brewery
6 Blundell Street
London N7 9BH

Senior Editor: Susanna Clarke

Project Editor: Charyn Jones
Consultant: Mary Trewby
Designer: John Grain

Illustrator: Charlotte Wess

Picture Researchers: Shona Wood, Sheila Geraghty

Art Director: Moira Clinch
Editorial Director: Carolyn King

Typeset by Ampersand Typesetting (Bournemouth) Ltd
Manufactured in Hong Kong by
Regent Publishing Services Ltd
Printed by Leefung Asco Printers Ltd, Hong Kong

16   15   14   13   12   11   10   9   8   7   6   5   4   3   2   1

Library of Congress Cataloging-in-Publication Data
Rees, Yvonne
   Floor style.
   1. Flooring. I. Herbert, Tony. II. Title.
TH2525.R44 1989   690'.16   89-5787
ISBN 0-442-23923-8

CONTENTS

◇

THE ROOMS

Chapter One

THE ART OF FLOORS

CLASSIC STYLES OF THE PAST

The dramatic impact and strong visual contrast of alternate checkerboard black and white marble tiles has remained a firm favorite for grand entrance halls for centuries. They may be originals, uprooted and transplanted to a new home, or new cut tiles – almost indistinguishable from the old – or recently, even be a total illusion, the effect of marble reproduced in high-quality vinyl.

GRAND FRENCH ENTRANCE

Many in-situ marble floored halls have survived intact to thrill the visitor, none more so than this high-ceilinged entrance to the Chateau at Anet in France. The eye-catching monochromatic regular pattern brings the floor into focus, insuring it is in no way dwarfed by grand stonework, ornate double doors and full-size guardians in armor. They hold the attention, yet, being totally devoid of color remain relatively neutral and therefore undemanding, and this is perhaps the reason for their endurance as a design theme. The owners are unlikely to tire of the effect and be tempted to change the floor for something more fashionable. Nor is marble likely to require replacing through wear, because of its very durability.

France always has led the field in floor tile design. Marble and painted tile ideas were being borrowed and transposed to ecclesiastical English floors and grand palaces as long ago as the thirteenth century; the introduction of decorated tiles expanded the scope for design to circular patterns and elaborate tile pavements complete with heraldic motifs, trefoils, crosses, latticework and borders using squares and rectangles. These would be in patterns of red, yellow, golden brown and olive green – combinations which survived until the nineteenth century and which are still to be found in many Victorian porch and passageway designs.

Around the seventeenth century prominent houses were beginning to adopt another perennially popular floor style for hallways and entrances. Stone or light

Previous page The entrance hall of the Chateau at Anet in France is a real eye-catcher with its classic floor of black and white marble slabs offsetting a rather grand and imposing interior. This style floor has been popular for all types of property for centuries.

marble slabs, usually laid on the diagonal, had cut corners inlaid with small squares of black marble at the intersection of every four meeting slabs. By the late nineteenth century more modest households could afford tessellated floors of black and white marble or colored stone, and again it was a style that was to remain with us long enough to embrace new materials and new techniques without changing the basic look.

ROMAN TROMPE L'OEIL

Incredibly fresh and witty after 2000 years, the mosaic floors of the ancient world prove their remarkable endurance by surviving fire, flood and volcano, to delight the curious modern visitor and inspire today's ceramic, stone and marble designers. "The unswept room" at Pergamon is a particular favorite, designed by an artist called Sosus and much copied around Rome in his lifetime. It was certainly admired by Pliny, while another famous Sosus mosaic, showing doves drinking from a cantharus or wine vessel, was reproduced for Hadrian's villa. Sosus' work is mosaic at its very best.

The "unswept floor" is in fact a clever and convincing *trompe l'oeil* portraying the kind of debris to be expected from the Roman dinner table; fruit seeds, peel, nut husks, bones, even scavenging mice, all depicted in tiny mosaic chips. This realism, a desire to represent nature in art, was a continuing theme of the emblema, or central panels of classical mosaics, undertaken by skilled artists and set into locally made mosaic floors. These might illustrate farm scenes or rugged landscapes, birds or fish, figures and buildings, or represent familiar tales of the gods and their exploits. The picture panels would be surrounded by carefully laid out geometric patterns using minute pieces of glass, stone or fired clay.

They are not simply decorative and entertaining to study. These early mosaics, found on sites all around the Mediterranean, tell us as much about the artists who designed them and the lifestyle of those who commissioned and lived with them, as cave paintings do about their inhabitants. No doubt modern examples are equally a rich legacy for future archaeologists and social historians, as they are a statement of the homeowner's wit, taste and lifestyle.

The "unswept floor" is a perfect example of Roman mosaic *trompe l'oeil* and was created by the artist Sosus at Pergamon. It realistically depicts what must have been everyday Roman litter using tiny fragments of colored stones and clay.

CASTLED SPLENDOR

The magnificent polished wood floor in Henry VIII's Banqueting Hall at Leeds Castle, Kent, England, runs a breathtaking 75ft beneath an elaborately carved oak ceiling. The floor in this major west-facing room is completely constructed in ebony; the irregular, narrow planks are beautifully joined, and the wear and tear of centuries produces an extraordinary variety of color and tone in the wood.

The Norman Banqueting Hall is on the second floor with splendid views of the surrounding lakes and grounds through a sixteenth-century bow window – evidence of Henry VIII's original rebuilding work. It was in the manor houses, halls and castles of the great and wealthy that wood floors were first used. When upper floors began to be inserted in new or existing buildings, their owners were looking for something lighter, warmer and more stylish than the stone flags or dirt and rushes they were used to.

The Henry VIII Banqueting Hall at Leeds Castle in Kent, England, features a magnificent ebony wood floor, running along the western side of this Norman fortification.

The sturdy boards would be hand-sawn as they came, so they tended to be of uneven widths and depths; often the supporting joists would be packed out to help them lie well. It was quite usual for the wood to serve as both floor and ceiling, the underside painted, later plastered, for decoration. Originally the boards were not fixed so that, if transferring homes, they could be removed and taken away like all the other furniture and furnishings.

Oak was more commonly chosen for early floorboards, but elm was also used extensively; when matured, it produces a fine mellow finish. With boards at least 1in thick and often more, lovingly waxed and polished, or hidden for hundreds of years under a succession of more fashionable floor coverings, many such floors have resisted rot and worm and retain their original glory.

It wasn't until the eighteenth century that ground floors became boarded, too. Carpets began to be used on the floor at about the same time – although, more commonly, painted oil cloths would be seen. Prior to that, rugs and tapestries were hung on the wall as here in Leeds, and only rush mats were used to soften stone or wood floors.

Existing wood floors in old houses should be restored wherever possible, and it is often feasible to salvage quite badly damaged boards where they are thick; once seasoned they can be remarkably resilient even to fire and structural damage. Sections can be cut out and new pieces put in where necessary, on the advice of an historical wood specialist.

Alternatively, it is possible to replace a floor with reclaimed boards or new wood: not ebony, but oak, elm or yew, perhaps pine for a traditional effect in nineteenth-century houses.

ELEGANT EIGHTEENTH-CENTURY SALOON

Exotically patterned woven carpets had been hung on walls and used to cover seats and tables in the west since at least the seventeenth century, but it wasn't until the late 1700s that they began to be seen on the floors of the grander houses and residences of the wealthy. With the rebirth of classical elegance, large country houses like Saltram House in Plymouth, Devon, England, were

The superb Axminster carpet in
the saloon at Saltram in Devon,
England, displays Robert Adam's
talent for co-ordination and design.
He planned this room from floor to
ceiling in 1768.

determined to exhibit the latest fashion to its fullest and finest extent, especially in the saloon or "great room" of the house, reserved for county balls, concerts and other entertainments and therefore designed to impress.

Robert Adam was commissioned to handle the design of the saloon at Saltram by Lord Boringdon in 1768 and it bears all the neo-classical Adam hallmarks: an elaborate plasterwork ceiling, attributed to Joseph Rose; carefully positioned mirrors and paintings (Adam naturally offered advice on subject matters for these); meticulously planned fixtures and fittings, right down to the door handles, valances and pilasters. Even the Chippendale chairs, tables and *torchères* were specially chosen and positioned to give prominence to the carpet, a magnificent woven Axminster, designed by Adam to echo the pattern of the ceiling but using stronger colors to

The splendid mosaic patterns and borders on the floor of the Arab Hall at Leighton House in London were specially designed by George Aitchison and were constructed exclusively of local marble.

complement the pictures on the walls. The carpet is truly magnificent, almost three-dimensional in detail and woven by Thomas Whitty at Axminster, for which he was paid £126 in October 1770.

Adam's close attention to detail, his meticulous balance of patterns and colors between every decorative feature in the room, paid off in the final effect and made him justly famous.

As a potential surface for design, the floor was, for him, as important as the walls, the heavily decorated ceiling, the fireplace, window dressing and furniture. This splendid overview is only recently being applied again to floors when planning both grand and more modest interiors. Rugs are being specially commissioned to create focal points, or carpets designed and fitted with custom-made borders and panels to outline the shape of the room or echo other decorative and architectural features.

ARTIST'S MARBLED MOSAIC

Masterminded by George Aitchison for Lord Leighton, President of the Royal Academy, Leighton House in London is a monument to some of the best in Victorian art and design. Aitchison's mosaic floor in the Arab Hall is particularly stunning. Completed around 1879-80 by Italian craftsmen but using local English marbles, it has all the finesse and detail of an exquisite oriental silk carpet incorporating specially designed borders and panels with a floral theme in subtle green/gray and creamy veined natural marble shades.

This classic marble "pavement" is the perfect background for a grand London house entrance, complete with lofty columns, a handsome staircase, even an indoor pool and fountain; Leighton House was an indulgent fancy of its day.

The Arab Hall mosaic floor still stands as a unique work of art, probably one of the finest to be seen in Britain. More common for nineteenth-century entrance halls were simpler patterns, circular and bordered designs devised using black and white marble chips with colored stones. These can be recreated today producing new sophisticated patterns to co-ordinate with contemporary furnishings and decorations, or be based on a traditional pattern within a period home.

The number of mosaic floors that have survived intact, not just in the grander houses but in humbler Victorian entrance halls, porches, passages and along front paths, is testament to the toughness and longevity of the material. An existing design could be difficult to repair if the materials or colors have become rare. Under these circumstances, it is a good idea to keep an eye open for fragments in architectural salvage suppliers, or to keep in touch with neighbors owning similar houses who may be digging up their own incomplete mosaics to replace with something new.

MANOR HOUSE KITCHEN

The traditional kitchens at Canons Ashby House in Daventry, Northamptonshire, England, have been barely altered since the manor house was renovated in 1710. Here is the classic stone-tiled kitchen floor: local stone, wherever suitable, would be used in large slabs or smaller sections with matching or staggered joints, sometimes in geometric patterns if the stone was available in regular shapes and sizes. The stone was generally laid directly onto bare earth – indeed, beaten or baked soil was often the floor being replaced. Later, in the early eighteenth century, thin brick or tile type pavers, kiln-fired from local clay, might also be used as a sturdy, practical floor.

In the Canons Ashby kitchen, the floor is still serviceable, despite being cracked and uneven. It was certainly a practical option when new, required to withstand running water from above and below and dropped food debris, as well as the dirt generated by a large double-fronted range in the enormous fireplace. The wood duckboards close by the sinks were often made of pine and were designed to keep the maids' feet a little drier.

Where existing stone flags and tiles are uneven and badly worn, with mold seeping through the joints or even through the slabs themselves, if they have worn very thin, the best approach is to lift and store them carefully, to put in a proper mold-proof course and re-lay the floor. It helps to have numbered the pieces when fitting the puzzle back together again. Worn slabs can sometimes be turned and used on the other side. Polish build-up can be cleaned off with washing soda, and for convenience when clean, they should be sealed with a brand name floor sealant to make care and maintenance easier. Some people successfully use a polishing machine, although this is not advisable where a slippery surface can be dangerous.

It is still possible to buy reclaimed flags from old homes, or to purchase used pavement slabs. These may be very uneven and some can be quite thick, which can be a problem for inside use where the top surface needs to be reasonably level. Shot blasting is one solution here. Slabs will have to be juggled around to get them to fit. Rent an angle grinder to cut them; a diamond edge does the job quicker and easier.

For traditional-style kitchens, stone is still a highly practical, good-looking choice, providing it is well laid. Newly quarried and dressed stone is still available for creating authentic or original effects; depending on locality they range from honey-colored York slabs to grays, buffs and sandstones, even imported stone tiles complete with fascinating fossils. Slate was traditionally used and is still available but needs regular treatment with linseed oil and white spirit to keep it in really good condition.

SHAKER SIMPLICITY

Clean simplicity and honest use of natural materials typify the Shaker interior, a look which seems to be enjoying a revival in the late twentieth century, but one which was never far away from modern thinking and design. The Shakers were a religious communitarian sect, which originated in Britain but settled in America in the nineteenth century. In pursuit of self sufficiency, members of the sect began to design and sell furniture, mindful of their dedication to functionalism and with strict adherence to the Shaker creed that "beauty rests on utility." The Shakers believed that to work hard for something that couldn't be put to practical use was sinful. The resulting inexpensive but extremely well-made furniture was simple and elegant, traditional designs distilled into clean lines – their ladderback chair is a famous example. In many ways the classic mini-malism of modern furniture has its roots here and in recent years Shaker style has gained in popularity.

The kind of room in which this furniture looks at home is equally beautiful in its spareness and simplicity:

The original kitchen at Canons Ashby House in Daventry, Northamptonshire, England, still has its traditional eighteenth-century stone floor, complete with wooden duckboards to keep the feet dry.

Above Leaving the wood floor plain suited early American austerity and practicality. The natural sheen also reflected valuable light into the room.

plain white linen, color- or white-washed walls, bare stone and stripped woodwork.

The Shaker style is very close to the much earlier North American Quaker style in many ways, only less decorated, with plain boards on the floors, scrubbed but not polished, and made of whatever wood was at hand, frequently pine. Rugs and carpets were certainly not judged necessary and were rarely seen; the glow of the scrubbed wood was used as a device to reflect available light from the rather small windows. The boards were often wide and thick, in upper rooms providing both floor above and ceiling for the room below – a dual functionalism that would have no doubt pleased the Shakers. The boards were plain, but like the furniture, the gleam

and grain of the wood was attractive enough to stand unadorned.

In this Shaker meeting room the excellence of workmanship lets the raw materials speak for themselves, creating a calm and lovely interior washed with light and the natural gleam of plain paint and wood.

The boards would have been regularly scrubbed with water or sand to keep them spotlessly clean and gleaming. These days, the same effort can be reproduced with far less effort by sealing stripped wood with mat or satin varnish which produces a much softer glow than shiny gloss. For a more decorative look, but closer to the Pennsylvanian tradition than Shaker philosophy, the floor could be painted verdigris green or ox-blood red.

Above Old-fashioned boards tend to be wide and of uneven sizes – taken from the tree as they come. Plain-scrubbed, they have a natural appeal like the simplicity and classic styling of traditional Shaker furniture.

In the nineteenth-century style conservatory at Dunster Castle in Somerset, England, colored borders and terracotta tiles were used to emphasize its length. The natural colors of the tiles blend well with plant material.

VICTORIAN CONSERVATORY

With the development of cheaper, more efficient glass in the nineteenth century, the splendid Victorian conservatory or glasshouse was born, a boon to the architect only too willing to create for his customers elaborate palaces of space and light and an essential addition to the house for those who could afford it.

If the roof shape was the pinnacle of conservatory design, the floor was its echo and anchor feature. Highly decorative in stone or colored tiles, it was a celebration of patterns and borders, following the shape of the main structure and perfectly reflected in the glass above it. Sturdy materials such as these were practical enough to cope with the Victorians' fervor for exotic plants, transforming their conservatories into mini-jungles. The conservatories were easily swept of soil and other stray plant debris, and could be hosed with water daily. Yet tiles were also highly decorative, despite their tough practical properties, satisfying the need to convert the glasshouse into an exotic living room by being as colorful as the real living room carpet.

Dunster Castle, near Minehead in England, dates back to the thirteenth century but was remodelled by Anthony Salvin in the 1800s. This classic Victorian conservatory maximizes the impact of a narrow terrace design with an arched roof structure and elaborate border of terracotta tiles on the floor. The mellow, earthy colors of the terracotta patterns are a splendid foil for raised beds and large matching terracotta plant pots. Planting is restricted mainly to exotic foliage plants with the occasional touch of colored blooms, so as not to detract from the subtly shaded floor design.

Here a strong border defines the shape of the conservatory and follows the line of the roof; traditionally a square, hexagonal or circular structure would center its floor design, using stone, ceramic, terracotta, or sometimes marble tiles to create decorative patterns, panels and borders radiating outwards to emphasize and reinforce the architectural design. The success of this technique and the popularity of this style has remained virtually unchanged to the present day in both interior and exterior conservatory design.

HAM HOUSE

This large, formal dining room is part of seventeenth-century Ham House in Richmond, England, a boat ride down the Thames from London's Westminster Bridge. Originally it contained a black-and-white tessellated marble floor but this was removed in the eighteenth century and replaced with a wood floor. It's a wood floor of some magnificence: a grand parquet design using eight different woods arranged in an extensive geometric design of squares and lozenges, as rich in texture and variety as any Persian carpet.

The subtle natural shades of the wood range from buffs and browns to russets and ochers and have been expertly blended, light against dark, to produce a wonderful three-dimensional effect. The warmth and grain of the wood enhances the pattern with its textures. Testament to wood's excellent durability as a flooring material, the wear and tear of the centuries has only improved and matured the surface, buffing it to a mellow glow of complementary colors.

This juxtaposition of light and dark, the eight woods with their variety of grain and color arranged in intersecting shapes, successfully disguises the true shape and size of what is an extremely large room, for, despite its spaciousness, it now has a cozy, friendly atmosphere. Because the colors are so natural and muted, the floor never overpowers or becomes too dominant: with polished wood panelling and similar earthy shades for other furnishings in the room, a perfect harmony of colors and textures has been achieved.

A clever combination of dark and light shades confounds the eye and can create a subtle, stunning floor. It is a technique that can be adapted successfully to more modest interiors. Such a splendid and elaborate arrangement of woods may be out of the question today, but a similar effect could be achieved in smaller rooms using a less ambitious parquetry design if you choose your woods carefully. Alternatively, try out your ideas in different shades of cork tiles, or imitation wood vinyl tiles – much easier to cut and fit and very effective. The idea could be brought right up to date in clear pastel linoleum or vinyl, blending the softest modern shades into graphic designs. A piece of graph paper should help to plot the design to scale and to work out exactly how much material is

Left The floor in the marbled dining room at Ham House near London is a magnificent cut and shaped design of natural grainy colors that produces an almost three-dimensional effect.

Right A breath-taking vista in the Ducal Palace, Valencia, Spain, where simple blue and white ceramic tiles have been laid to emphasize the perspective of a series of grand ante-chambers.

required – rather in the same way a patchwork quilt might be planned. You only have to look at log cabin and some of the other, more intricate patchwork designs to see the similarity in approach.

It is possible to fake wood grain effects using paint and/or a special scumble glaze, available from specialist decorators' suppliers, and floors exactly like the one at Ham House could be created on a plain surface such as hardboard, laid smooth side up. However, paint graining to this standard over such a large area would be extremely time-consuming and many coats of varnish would be required to protect the design.

SPANISH PALACE

A repeated geometric design in limited colors – from the classic black and white marble of the entrance hall to the gay red and white kitchen tiles of the 1950s – has always offered an excellent opportunity to play games with perspective. What better choice to draw the eye along a breathtaking series of ante-chambers, through spectacular, lofty, decorated doorways, than the simple blue and white ceramic tiled floor in the Ducal Palace, Valencia.

It is the kind of floor you might see in the simplest Spanish hacienda: plain blue and white tiles arranged alternately in the familiar checkerboard design. By positioning the tiles on the diagonal, those plain squares are converted to the sharper profile of a series of diamonds or flattened lozenges. Run the design straight along the passage so attention is automatically riveted towards its furthest reach and you have created a vista worthy of the grandest paintings.

Cool, crisp and uncompromising, the floor has become the perfect foil for its grand surroundings, strong enough to balance the painted ceiling, sufficiently simple not to compete with elaborate decorative details and palatial grandeur. In the Ducal Palace, the floor is designed to stun the senses and humble the spirit, to take the breath away with the scale of it all.

Similar magic can be worked to a smaller extent in the home by laying alternate tiles diagonally across a hall or bathroom floor to disguise its size. Alternatively, plan your design to make a room look longer or wider than it really is by laying the pattern in the direction you wish to extend visually.

Chapter Two

MATERIAL POSSIBILITIES

◇

BEGINNING WITH BASICS

The floor is the most worn surface in the house. It has to stand up to a daily pattern of use that would wear some materials to shreds in hours. Fortunately the range of natural and synthetic floor materials has never been wider. However, a new floor only looks good and remains in good condition if it is well laid, that is, with a professionally laid level sub-floor or the appropriate underlay.

Some floor materials like carpet, vinyl and ceramic tiles are simply floor coverings – they are not floors in a structural sense. Concrete and wood, however, can be both floor finishes and structural building materials.

The term "tile" may be confusing. Sometimes it refers specifically to ceramic tiles, although in practice most floor materials are available in tile form. Tile simply means a modular form of the material, often, but not always, square, which has the advantage of being easy to transport and handle and allows the possibility of chessboard and other patterns when laid in contrasting colors.

NATURAL MATERIALS

Stone or rock are generalized terms for almost any hard natural material from the earth's crust. In all their great geological variety they construct the natural scenery of the world. From the earliest time we have used them in a similar fashion to build houses and to lay floors.

Stone As a flooring material stone is invariably hard, cold and unyielding, but there are great variations in color, texture and hardness. These differences are a product of the different components of the stone, the way they are held together and their geological origin. The different rocks and stones are made up of minerals – substances of a definite chemical composition. Thus granite is chiefly made of three distinct minerals: quartz, feldspar and mica.

Some rocks like sandstone and limestone have more or less distinct parallel layers or strata. They were formed by the hardening of layers of sediments accumulated at the bottom or on the shores of a sea or lake, or heaped

Previous page Shantinatha tiles at the temple site of Khajuraho in Madhya Pradesh in north central India. At one time fifty Hindu and Jain temples, dating from the tenth century, stood at Khajuraho, of which twenty-two have survived in good condition.

Right Sawn and polished slabs of English limestone in a random arrangement: the dark blue color is punctuated with an intriguing array of fossil remains which highlight its organic origins.

Below Smooth natural sandstone sawn into precise geometric shapes offers an enduring floor. Subtle color variations are due to natural impurities in the rock. Sandstones such as these require sealing to insure that their light colors are retained.

together by the wind. For this reason they are known as sedimentary or bedded rocks.

Other rocks such as granite are more crystalline in appearance. They show no trace of a bedded structure and were formed by the cooling and hardening of molten material forced up from within the earth's crust. They are known as igneous (of fire) rocks. Thus, its hard-wearing qualities apart, a natural stone floor offers a fascinating piece of the earth's history, probably millions of years old. Some rocks also contain fossils, the hardened remains or traces of long-extinct animals and plants.

Traditionally stone floors and buildings would have been made from the nearest available suitable stone. Thus the architectural character of an area is often a creative human expression of the local geology. The warm, honey-colored stone cottages of the English Cotswolds are a direct extension of the limestone rocks on which they stand. Today stone is transported with comparative ease, introducing materials from exotic places to bring their own contrasting character.

Rough-hewn natural stone straight from the quarry will look very different from the same material when ground and polished. Usually polishing brings out the "texture" of the rock, even though the surface texture is smooth, and makes it more colorful. Modern sawing techniques have made stone more widely available for domestic use. Thin sliced stone is cheaper and lighter to transport and install.

As well as on outdoor terraces and patios, stone is a good flooring for porches and conservatories, kitchens and hallways. This depends on the climate; some stone isn't suitable as outdoor flooring. Its cool ruggedness contrasts brilliantly with colorful rugs which can be thrown down for warmth and comfort. The durability of stone is legendary, but the more porous types such as sandstone require sealing to avoid the possibility of unsightly stains.

Limestone The limestones of different regions vary greatly in color and hardness and may be almost white, creamy, gray, browny-red or even nearly black. They consist of a high proportion of the mineral calcite, a form

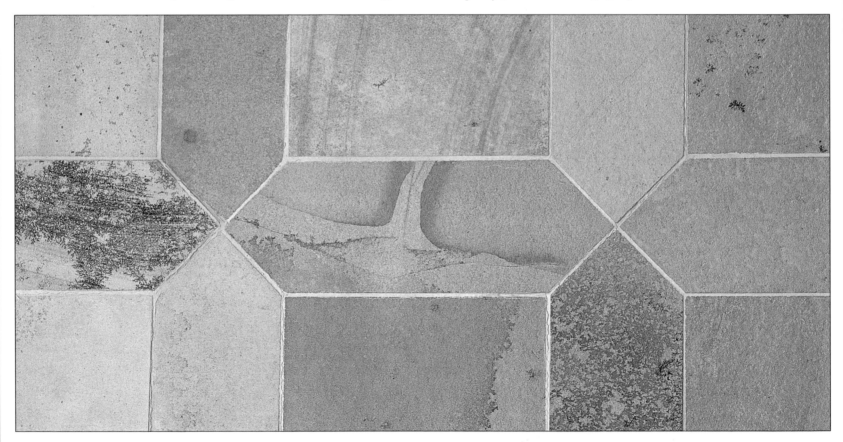

of calcium carbonate, which is abundant in the earth's crust. Some of the most attractive flooring limestones appear to consist almost completely of fossils and when cut and polished give a fascinating cross-section of the corals, sea lilies and shellfish from a warm tropical sea of millions of years ago. Just like a heavily patterned carpet, they can be overwhelming, and their use is often most successful as contrasting bands in other, plainer stone.

Travertine is another form of limestone deposited from springs. It is a porous light-yellow rock which hardens on exposure to air. It has had a long history of use in Italian buildings and provides a floor of much greater textural interest than smooth marble.

Sandstone As its name implies, this is a stone made of grains of sand cemented together with silica. The sand grains consist of quartz, a form of silica, which constitutes over a quarter of the earth's crust. It makes a strong and good-looking building stone with attractive shades of cream, pink, red and, rarely, green.

The grains in some sandstones are large and angular and these are known as gritstones. The texture of mill-stone grit was admirably suited to the grinding of corn and today, although millstone grit is not widely used for flooring, old dressed circular millstones can make a striking historical feature in a floor.

Sandstone on the floor can feel rough and abrasive, like walking on sandpaper. More comfortable are the fine-grained sandstones or siltstones. The grains may form thin layers along which the stone easily splits. This is the source of the familiar stone slabs or flags of old kitchen floors that have a strong nostalgic appeal.

Square or rectangular slabs of stone, even in different sizes, give a disciplined, modular feel to a floor. Using irregular stones, sometimes referred to as crazy paving, needs to be done with care to avoid a confusing appearance. Many types of stone are now crushed and reconstituted to manufacture units of flooring. While cheaper than natural stone, they may have a bland appearance and artificial textures can look contrived.

Cobbles or pebbles These are smooth, rounded stones of a variety of different rocks. They have been formed from irregular lumps by the tidal movements of the sea or by the grinding action of ice in glaciers. They are an

ancient and inherently outdoor form of stone flooring, much used around the Mediterranean, and they can look striking. The smoothness of the individual stones contrasts with the rough texture when laid as a floor, and they present quite a challenge to bare feet!

Although large pebbles literally cover the ground faster than small, the latter give a more comfortable and visually satisfying floor. While cheap on materials, cobbles are labor-intensive to lay. They need to be bedded deep and carefully into sand or more permanently into weak cement mortar. White quartzite pebbles can be used to great effect to create bands, borders or patterns among darker cobbles.

Square or rectangular blocks of stone such as granite are sometimes known as cobbles, although they are more strictly stone sets. They are easier on the feet than pebbles and when laid in interlocking fan patterns make a stylish driveway that will withstand vehicular traffic.

Above left The success of this formal cobbling treatment relies on the careful choice of size and color of stone. Essentially an outdoor flooring, cobbling can be used with other types of paving, especially as a border.

Above This bathroom in the grand style makes extensive use of colored marble on most surfaces.

Marble Slate and marble are both examples of rocks which have been transformed in their structure and chemical composition by being subjected to intense heat or violent earth movements. Such rocks are known to geologists as metamorphic rocks, a term derived from the Greek word meaning "change of form."

To the ordinary person marble simply means "an ornamental stone that takes a polish" – an all-embracing term which would include polished limestones and travertine as well as serpentine. Strictly speaking marble is a limestone which has been altered and in which the calcite has been re-crystallized. The crystalline nature of the rock is the key to its dramatic brightness and sparkle

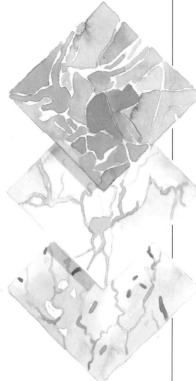

Left Colored marbles used in an intricate way can create an opulent interior. Marble looks and feels cool and is a perfect complement to the splashing water of a fountain.

Above Three examples of marble giving an indication of the wide variety of colors and patterns of veining that is available: from top to bottom the examples are Rosso Alicante, Calacata Gold Vein and Perlato.

when it is used in polished form on a floor. The purest form of marble is white but, in practice, chemical impurities such as oxides of iron are always present, and this explains the polychromatic veins and markings that are so characteristic of marble.

Italy is still renowned as the world center in marble and, although the material is quarried in a variety of other countries, a high proportion of the blocks are sent to Italy for processing. Contrasting black-and-white squares is a traditional use for marble, but with the huge range of colors available greater subtlety can be achieved.

Marble in the home looks grand but is cold and hard underfoot and extremely slippery when wet. Large sheets of marble are the most expensive form because many blocks will only yield small, broken pieces. These fragments are used for standard-sized marble tiles where a thin facing layer of marble is fused to epoxy-resin-impregnated glass fiber. The resin fills, seals and strengthens any natural cracks and fissures in the marble, and the glass fiber backing enables the tile to be transported and laid with relative ease.

Slate Although best known as a roofing material, slate makes hard-wearing and dramatic floors. The rock is characterized by the relative ease with which it can be riven (or split) along parallel planes of cleavage to produce thin sheets.

Although it will slice and polish well, the naturally riven surface has a gentle rippling texture which looks attractive and helps to prevent slipping. Colors vary

Below Slate is equally at home out of doors, whether on the roof or as a paving material. The natural riven surface of the gray slate on this garden terrace provides a delicate texture, catching oblique light and offering some resistance to slipping on wet conditions.

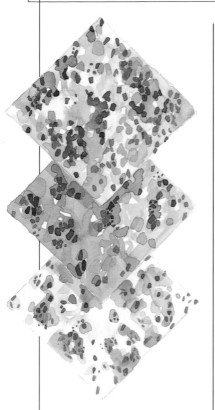

Above Three examples of granite. Here the variations in color and texture are more subtle, but there is still enough choice to enhance most decorative treatments.

Left Slate can be cut into rectangular modules and laid in a regular fashion like bricks or tiles. The natural color of the rock is enhanced and darkened by sealers and polish. The joints between the slate can be emphasized by the use of a light mortar, as here.

according to their source. Spanish slates are light gray through to the deepest midnight blue. North Wales slate is a deep gray/blue, occasionally purple, while green slate is available from quarries in the English Lake District. Flooring slate exported from Africa has rich brown and red colorings and is available in square tile form.

Granite This has a large-grained homogeneous texture providing a denser, harder material than marble. The main centers for quarrying are in Africa, Scandinavia, Italy, India and Spain. Much is sent to Italy for processing because of the great expertise the Italians have built up dealing with these hard materials.

Granite, for a domestic floor, needs to be used with care. It is expensive, hard and, as it is frequently adopted for the most prestigious commercial and public buildings, may create an over-opulent image in the home.

MANUFACTURED MATERIALS

The quest for floors with the hardness of stone but without the need for joints has brought about the use of synthetic floor materials and techniques such as concrete, terrazzo and mosaics. The continuous, jointless quality of these heavy types of floors has important practical advantages such as cleanliness and hygiene. They do, however, depend on sturdy sub-floors and for this reason the domestic use of these materials is usually restricted to solid floors at ground level. In other situations, load-bearing and rigidity need to be checked. There are few flooring faults more unsightly than long wandering cracks in terrazzo and mosaic caused by floor movement.

The nineteenth century not only saw the development of sophisticated machinery for processing natural floor materials but also an advancement in chemical knowledge which literally paved the way for a new and important family of synthetic floor materials. The trend towards rolls of flooring, each covering large areas and therefore having the minimum of joints, was set by linoleum and followed by rubber. Today vinyl is a generic term for a large commercial range of flooring based on the complex plastic polymer known as polyvinyl chloride – PVC.

Concrete This surprisingly ancient synthetic building material is an intimate mixture of water, an aggregate of sand and stones, and a binder or cement which hardens to a stonelike mass. Most parts of the world where early civilizations were established had natural cement deposits. The Romans, for instance, made concrete from a fine chocolate-red volcanic sand or pozzolana. In the twentieth century concrete has become a dominant material in the construction industry. The cement used in today's buildings originates from the pioneering work of Joseph Aspdin who, in England in 1824, patented a cement made by firing a slurry of clay with limestone and then grinding the residue to a fine powder. It became

Above Traditional mosaic patterns.

Above left Terrazzo tiles often resemble marble and share similar qualities of hardness and coolness. The polished finish brings out the color of the constituent stones. It provides an easily cleaned, waterproof floor – ideal for a bathroom.

known as Portland cement because it resembled a fine English building stone of that name which was used for St Paul's Cathedral, London.

Despite the implied associations with natural stone, Portland cement is, for many people, a dull and mundane material, especially when used for floors. Its value lies primarily in its use as a structural and sub-floor material where it may be reinforced with steel rods. As the work horse of flooring materials it is very much at home in utilitarian spaces such as garages, driveways and cellars. It can be left rough or ground and sealed. Cement paints are available which can bring color to an otherwise heavy gray mass of floor.

Terrazzo Sometimes known as Venetian mosaic, terrazzo consists of colored stones, laid in cement mortar, in a layer about ½in thick over level concrete. When the mortar has set, the stones are ground smooth by machine. This brings out the color of the stones and provides a much brighter and more interesting floor than plain

concrete. When used over a large area, narrow strips of expansion joint material are inserted every few yards to minimize cracking. Modern epoxy resins are increasingly being used as the bedding material for their hard-wearing and chemical-resistant properties.

A terrazzo floor is fairly expensive to install. It can be a bold, simple design using different colored stones, and the grinding-off process characteristic of terrazzo insures an even and easily cleaned floor, suitable for a modern kitchen.

Mosaic Fine mosaic floors from excavated sites throughout the former Roman Empire bear witness to the early and sophisticated development of this decorative form of flooring. The Romans used a repertoire of standard motifs which, combined with a high degree of detail and good use of color, have been admired and widely copied.

The fundamental basis of mosaic work (which can be applied to walls as well as floors) is the combination of different-colored small pieces of hard substances like

marble, ceramic or glass to form a design that may be geometrical or pictorial.

At its most delicate, mosaic is almost like painting with small solid pieces rather than pigment on a brush. The small pieces are known as tesserae and their assembly into pattern or picture is the labor-intensive and therefore costly aspect of mosaic work.

Different mosaic materials produce different effects. Marble can give great subtlety and shading. The range of colors available is enormous and, being natural, they combine well together. Within each individual tessera there will be a mottle or gradation of color. Thus marble mosaic has many of the qualities of a marble floor and can look soft and natural. Ceramic bodies can be colored with oxides and other pigments to produce a wide palette of raw material for mosaic floors. Unlike marble, how-ever, the color runs evenly through the body and as a result ceramic mosaic may look flat and even. Glass will provide the most colorful material of all, and some effects in mosaic work rely on the jewel-like brightness of

Above left Two examples of the almost infinite number of mosaic patterns and borders – these can be built up using geometric shapes such as squares and triangles.

Above right This new mosaic pavement is built up entirely from small squares.

Right To make a mosaic, the small pieces (tesserae) are cut or trimmed before the appropriate colors are pasted down onto a full-size drawing of the design. This is inverted and set into wet mortar in its final position. The thick and thin tesserae will all appear at the same level.

small pieces. In this way colors such as gold, rich ruby and bright peacock blue can be attained.

The introduction of bathrooms into homes at the end of the nineteenth century opened up a large market for floor mosaics which has since been extended to saunas

Left An almost three-dimensional effect is created by this skilful composition of shaped and cut linoleum. Far greater versatility is now offered by synthetic materials, such as vinyl and linoleum, which come in strip form as well as sheets.

Below Black, studded rubber lends itself well to this domestic kitchen. Generally thought of as an industrial material, it is a sturdy alternative for the home. For bathrooms, kitchens and utility areas its continuous surface with welded joints is easy to clean and comfortable to walk on.

and swimming pools. There is no need to place each tessera individually and laboriously into the floor! Pieces are stuck, face down, onto a full-size drawing of the design. This is then turned over and placed onto wet mortar or fixed to a prepared screed with adhesive. When grouted, the resulting floor has a good anti-slip property, useful for passages and steps, and shares the aesthetic qualities enjoyed by the Romans 2000 years ago.

Although geometric designs in mosaic are traditional and look good, abstract designs in modern homes offer great potential. By adopting square mosaic pieces it is possible to use computers to generate sophisticated designs including figurative elements and monograms that hitherto have been difficult and expensive to produce.

Linoleum Despite its 1930s bathroom image, linoleum is older than many people imagine. It was patented in 1860 by an Englishman called Frederick Walton, and many of the early designs were garish printed copies of carpets with heavy floral motifs. One of the characteristics of linoleum is that the hard-wearing surface material is pressed onto a fabric backing traditionally made of jute. Until quite recently linoleum has been out of fashion and its one-time image of a brittle material, cracking easily, has been difficult to dispel. Today it is made from linseed oil, ground cork, wood-flour and resins, baked slowly at high temperatures and then pressed into a jute or fiberglass backing. The resulting material is thicker and much more pliable and durable than its old-fashioned counterpart. It has been used with imagination by contract interior designers (as indeed have carpets) and the domestic possibilities are now being realized. The reception areas of several large public companies and the exhibition halls of many contemporary museums boast linoleum floors of great style.

Linoleum is fairly warm and resilient underfoot and is hard-wearing. Care must be taken when cleaning to avoid water getting under the edges since this will readily rot the jute backing.

Rubber Natural rubber comes from the milky latex of tropical rubber trees and tends to soften in the heat and turn brittle in the cold. Once used almost wholly for rubbing out pencil marks, the natural material began a new industrial life thanks to the invention of Charles

Goodyear. He patented his method of "vulcanizing" the latex by heating it under pressure with sulfur. Rubber tires for motor vehicles led naturally to thoughts of rubber floors. These were first used in public buildings just after the turn of the century, and pioneering examples were installed in Guy's Hospital, London and by Pirelli at Rome train station.

Since the 1940s it has been possible to make synthetic rubber. This forms the basis for modern rubber flooring which is now widely used in airports and shops, operating theaters and for the floor space surrounding indoor swimming pools. It is frequently textured with shaped raised studs, making it instantly recognizable. Studded rubber flooring offers both underfoot comfort and safety as well as easy maintenance and a high resistance to cigarette burns (unlike vinyl). It has a modern high-tech image and can work well in the home, for instance in bathrooms or utility areas. The color range is enormous. Any joints can be welded to give a really hygienic continuous floor covering.

Vinyl This versatile and thoroughly synthetic material owes nothing to nature and everything to the skills of late twentieth-century scientists. Yet this chemically sophis-

Below The sophistication in vinyl flooring is made possible by a wide spectrum of colored borders, strips and special cuts. These present the floor designer with a rich palette of visual elements.

Above left Vinyl is used to good effect in the bathroom where ease of cleaning and water resistance are essential practical qualities.

Left Different types of sheet vinyl can be combined to simulate a sophisticated floor. Here blockwood vinyl, looking like parquet, is surrounded with a broad inlaid border and stripwood, both in vinyl.

Above right Modern linoleum is used with startling effect at the Theatre Museum, Covent Garden in London. Different shapes and colors, together with perspective, are used to advantage. This floor dispels the old-fashioned bathroom image of the material.

ticated material lacks a distinct visual image. It is the chameleon of flooring, being able to take on, sometimes with great conviction, the appearance of almost every other type of floor. Thus it is possible to buy floors of marble, granite, wood, ceramic tile and brick all in vinyl – complete with texture where appropriate!

The amount of actual PVC in any particular vinyl flooring varies considerably from around a quarter to over three-quarters of the total constituents. PVC is expensive so high content will mean higher price but with greater wearing qualities. Sometimes there is a thin layer of clear PVC on the surface, while the thicker base layer has a lower PVC content combined with a mineral filler. To increase the non-slip properties, grains of quartz are embedded in the surface layer. Some vinyls have built-in underlays of PVC foam to increase sound and heat insulation and give the floor a pleasant resilience.

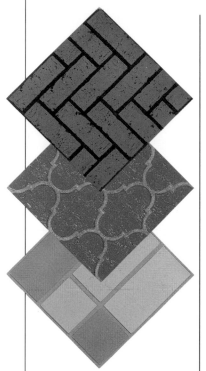

Above Traditional herringbone (top), French country style (middle) and modern-looking squares and rectangles (bottom) are just a few aspects of vinyl's chameleon character.

Right Rectangular clay tiles laid in a herringbone pattern provide a pleasing, simple floor.

Vinyl flooring comes in a bewildering range of patterns and colors and in both roll and tile form. To help break away from vinyl's safe middle-of-the-road image, some manufacturers are supplying plain colors in rectangular strips and other standard shapes, offering far greater scope for imaginative floor design.

TILES AND BRICKS

Tiles and bricks made from fired clay provide a synthetic extension to the range of natural flooring materials, such as stone, slate and marble, that are available from the earth's crust. Slithery, sticky clay seems an unlikely starting point for beautiful and hard-wearing floors. The fundamental change in the materials takes place when the clay is fired to high temperatures in a kiln. This was an early technological discovery, and consequently ceramic floors have a long and worldwide history.

Tiles The earliest ceramic tiles were made and used by the Ancient Egyptians 4000 years BC. Mesopotamia and China were other important early centers. The tile architecture of Persia, which developed during the Islamic period, has never been surpassed, with colorful ceramic tiles used to clothe walls and domes as well as the floors of buildings. In central and western Europe during the Middle Ages browny-red floor tiles inlaid with buff patterns became widespread. In England their manufacture and use was often associated with monasteries, and designs such as knights on horseback and fleurs-de-lis were common. Spain and Portugal developed lighter colored, hand-painted floor tiles well suited to their sunny climate. In the nineteenth century mechanical means of production allowed vast quantities of geometric and patterned ceramic floor tiles to be exported from Europe to America and Australia where many still bear the daily passage of feet.

Ceramic tiles at their simplest are nothing more than plain slabs of clay shaped in squares or rectangles. Traditionally the clay, of a consistency of plasticine, was hand-molded into wooden boxes and the resulting tile dried and then fired in a kiln. Today mechanical presses use almost dust-dry clay. This is more cost-effective and labor-saving than drying wet clay. These simple tiles are sometimes referred to as terracotta, meaning literally

"burnt earth." An alternative term is quarry tiles which has nothing to do with holes in the ground but is derived from the old French word *quarrel* meaning a square. Not surprisingly, the traditional qualities of these tiles are much sought after today; they have no pattern to wear out, only a patina of age to gain.

The joints between one tile and the next are almost as important to the look of the floor as the tiles themselves. If they are too wide, they can impose a dominant grid pattern even on plain tiles. With handmade "rustic" tiles, some irregularity of jointing is inevitable and is part of their attraction, but with crisp machine-made geometric shapes, precise jointing is essential. Some modern tiles are made with lugs (small projections from the side of the tile) to insure an exact fit to the surrounding tiles. Pigmented mortars can be used to color artificially the joints. For example, black works well with dark red tiles,

Left A crisp grid effect is achieved simply by the repeated use of three tiles of different shapes, sizes and colors. This arrangement would work as well on an outdoor patio or conservatory as inside the house.

Above Antique terracotta tiles from France: there is no pattern to wear away, only the patina of age to gain.

Right The juxtaposition of octagons and smaller squares can be used with a variety of flooring materials including marble. Here, handmade terracotta tiles create a country-cottage feeling. The same arrangement in glazed, machine-made tiles would look quite different.

making the joints look much less prominent and thus producing the "warmer" appearance of old floors.

Square tiles can be laid in straight rows in both directions, like the squares of a crossword puzzle. This works well in a square room, especially when used with a border. For corridors, it is worth considering offsetting or staggering the joints along the length of the space. This avoids the effect of long parallel rows of tiles and joints converging into infinity and optically makes a long narrow corridor appear shorter and wider.

In Europe in the middle of the nineteenth century there was a great revival of floors with patterned inlaid tiles, with increasingly elaborate designs using as many as eight colors. Such tiles are frequently known as encaustic tiles, meaning "burnt in" and referring to the way in which the inlaid colored clays are fired, along with the body of the tile, in one single operation. Herbert

Above A striking effect can be achieved by using black and white ceramic tiles set diagonally to the walls. The double border of brown tiles helps to define the space.

Left Large plain tiles of contrasting colors can be laid in panels with narrow borders to create a feature in a groundwork of smaller tiles.

Right These patterned tiles are formed by inlaying clays of different colors.

Left Magnificent floors can be built up from geometric tiles of different colors. Here in the Quebec Parliament Building an even richer effect has been achieved with the judicious use of regularly spaced patterned tiles.

Right In Europe there has been a tradition of using unglazed tiles in delicate subtle colors. Here the pattern is given "texture" by the use of speckled colored clay bodies.

Below Rustic charm is the predominant character of these simple terracotta tiles. Slight variations in the clay body and the tile size produce a warm-hearted, old-fashioned look.

Minton was one of the pioneers of this process, and Queen Victoria and Prince Albert chose encaustic floor tiles for Osborne House, their country home on Britain's Isle of Wight, completed in 1851. The prestige of royal approval was extended further by their use in the new Houses of Parliament, London and the Capitol in Washington, DC.

Today most patterned ceramic floor tiles are printed rather than inlaid with colored clays, and Italy and Spain are major centers of world production. The application of modern technology to traditional skills has meant that though tiles can now be thinner and flatter, you still need to specify floor tiles because wall tiles are not strong enough. Thin tiles help to reduce the overall weight of a floor – an important practical consideration when installing tiles on a suspended wood floor. As an alternative to fixing with cement mortar, a range of modern adhesives are now available which cope with particular problems such as damp areas and fixing to floors where there is a degree of flexibility and movement.

The flatness and precision of crisp factory-made tiles can be a virtue in a modern high-tech setting. For more traditional interiors, the handmade tiles from craft-based workshops, whose output may be small but highly individual, are a good choice. In some cases it may be possible to buy and re-use old tiles, but it is important that the edges are clean to allow neat joints when relaid.

Above Raw fired clay of a rough consistency produces the vernacular quality of these antique French terracotta tiles. The tiles are unglazed and to some extent owe their variation in color to their position in the kiln during firing.

Left Geometric tile patterns from a Victorian pattern book. Specialist firms can still be found to restore a floor like this or to produce tiles to create a new one.

An alternative to using patterned tiles to create a decorative floor is to build up a geometric pattern using a variety of shapes, such as triangles, diamonds and hexagons, in different colors such as red, black, buff, chocolate, blue, white or green. You could incorporate some patterned tiles to build up a Persian carpet effect. But beware! Too much pattern can "kill" a floor and simplicity is frequently the best solution. For a good, traditional-looking floor from inexpensive materials you cannot go wrong using black with either red, buff or white plain ceramic floor tiles laid alternately in diagonal rows.

Practical considerations when selecting ceramic floor tiles as against other materials must include their qualities of coldness and hardness. The former can be desirable in hotter climates and can be modified by the use of underfloor heating; but the latter is a fact of life, and if you drop your favorite china, it won't stand a chance!

Cleaning ceramic tiled floors does not present many problems. The tiles respond well to domestic floor cleaners and for really dirty ones, several manufacturers produce ranges of brand name tile cleaners. Ceramic tiles should last for years, but grit is the long-term enemy of patterned tiles. Over the years there will be general abrasion of the surface with eventual loss of pattern, caused particularly by the grinding action of footwear on grit particles. Frequent vacuum cleaning will reduce grit particles to a minimum.

Bricks Molded from clay and fired, bricks have provided convenient building and flooring units for thousands of years. Until the advent of modern transport, they were usually made as near as possible to where they were needed, so the warm reds, pinks and browns of old bricks reflected the local clay deposits from which they were made. Much more recently sand-lime or calcium silicate bricks have also been produced by compressing a mixture of damp aggregate with slaked lime and "steaming" it in an oven. They are pale in color (white, cream or soft pink) and so reflect more light than clay bricks. Pigments can also be added to produce more strident colors.

Brick floors are hard and robust and well suited to semi-outdoor areas like porches and conservatories as well as in the garden. Bricks can be laid on edge and in this way present a neat narrow appearance. However, they are usually laid flat and this is more economical, since they will cover a larger area than the same number of bricks on edge. Most modern bricks are formed from continuously extruded bars of clay which are cut with wires to make individual bricks. The four extruded faces are smooth but the two wire-cut flats are slightly rough, giving the brick when used for flooring an attractive texture which also increases the resistance to slipping.

It is easy to create a variety of patterns with brick. Herringbone looks good and is easy to lay. Bricks in straight rows can emphasize a sense of direction in a path, whereas basketweave produces a more multi-directional feeling suitable for patios and terraces.

Above Modern colored bricks in the historic setting of Kentwell Hall, Suffolk, England – the Tudor rose provides a tantalizing maze at ground level and a unifying element when viewed from above.

Right The wire-cut surface of these handmade bricks offers resistance to slipping. Sealing has removed the dusty appearance of brick floors indoors and will prevent staining.

Bricks vary widely in their density and porosity and should be chosen with care. The hardest and least porous are engineering bricks, traditionally used for culverts and railway viaducts and usually a rich slatey-blue in color. Although more expensive than ordinary bricks, their rugged industrial quality can be exploited to advantage.

Outdoors the growth of moss and algae on brick paving needs to be controlled, and smooth bricks can be slippery in the wet. Some bricks are produced specifically for flooring use and these are known as pavers. They are usually thinner than building bricks, may be of interlocking shapes, and can incorporate simple decorative patterns in relief such as diamonds and even Maltese crosses. This helps to prevent slipping, and for this reason they were traditionally used in stables and yards.

NATURAL WOOD

Wood is a natural and universal material that can be both the start and the finish for good floors. Suspended wooden floors are still common in the houses of many countries, particularly on upper stories. At their simplest they consist of wooden boards supported at regular intervals on beams (known as joists) which run at right angles to the boards. The resulting wood floor may simply provide the starting point for another type of floor finish, such as vinyl or wall-to-wall carpet, in which case the color and finish of the boards is of no consequence. In many cases, however, including wood on solid concrete floors, the wood will be the final decorative finish and its form, color and hardness will be important considerations.

To understand wood as a material, the qualities and limitations exerted by its origins as a living and growing tree must be appreciated. The trunk of the tree consists of millions of cells, the basic structural unit of plant material. In most trees the formation of wood cells follows an annual cycle with both growing and dormant seasons. This results in the visible concentric growth layers or annual rings which determine much of the character and quality of the wood. The rings vary in width according to the species and as a result of different growing conditions.

Water from the soil, carrying nutrients, enters the roots and moves upward through the wood cells to the

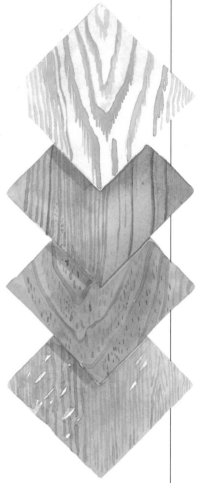

Above Four woods suitable for flooring. From top to bottom: ash, elm, beech and oak.

Left and right Narrow strip wood floors suit clean, modern interiors. "Secret nailing" avoids the need for visible fixings and allows the natural wood to display its full character. Pigmented beech strips (this page) look calm and thoroughly modern. The mellow honey tones of natural wood (right) provide a good setting for simple modern furniture. When used throughout, varying the direction of fixing the strips can help to distinguish different areas such as rooms and corridors.

leaves. The green leaves with their chlorophyll use the energy of the sun to combine the water with carbon dioxide from the air. In this way the leaves release oxygen to the air and produce a basic sugar for the tree's own use for building new wood cells. Thus it is important to remember that every piece of wood from a matchstick to the humble floorboard has been part of a living and growing system. It lacks the predictability and uniformity of a synthetic material, and part of its visual attraction when used in a floor is that no two pieces of wood are ever identical.

Not only does wood vary naturally, but its character and appearance is affected by the way it is sawn from the tree trunk. Wood for general and structural use is simply sawn in slices along the log. This method, known as "through and through," is simple and yields the maximum amount of usable lumber but is not selective of quality. For high-quality lumber, the log is sawn in quarter sections to obtain a cross-section of the tree which is stronger, more stable and of even appearance. However, this method leaves a certain amount of waste, making it more expensive. Another selective alternative is to saw at a series of tangents to the growth rings.

When a living tree is cut down, the wood is very wet because the cell structure is full of watery sap. As the water dries out the wood will partially shrink and this process, known as seasoning, can be done by allowing moisture to evaporate naturally or, much quicker and more controlled, in kilns. Eventually a fluctuating moisture balance between the dryness of the wood and the humidity of its environment will be reached. If this balance is upset drastically, shrinkage or swelling will result which can lead to warping, resulting from variable shrinkage in different directions.

Being an organic material, wood is susceptible to decay from fungi and wood-boring insects. Fortunately, many hardwoods are naturally resistant and modern softwoods are impregnated under pressure with preservative. This penetrates deep into the cellular structure of the wood and becomes permanently fixed within the cells.

For most people, all the disadvantages of wood, arising from its natural origins, are far outweighed by its qualities as a warm, hard-wearing and infinitely variable floor material. If you inherit really old wooden floor-

boards, they are probably of many different widths nailed directly into the joists with straightforward frankness. Despite the open joints between the boards and the slight relief provided by knots which have been more resistant to wear, old boards have a quality and charm of their own. Spare a thought for the enormous human effort involved in hand sawing a tree trunk to boards — hence the different widths.

In the nineteenth century the Industrial Revolution brought power and machinery to the business of sawing lumber, and even-width boards became common, usually with tongue-and-grooved edges, thereby providing a continuous, draft-free floor. In Europe, native hardwoods such as oak and elm were by the eighteenth century supplemented by fir (which when sawn is known as deal) imported from Scandinavia and the Baltic. Today a huge variety of over 150 different floor woods is available from all continents of the world. Modern wood boards or strips are usually "secret nailed" through their edges, thus giving a clean and neat appearance. As a complete contrast and to satisfy nostalgic demand, it is also possible to buy wooden boards supplied with black neoprene strips for the joints, recreating the appearance of a ship's decking of sturdy timbers caulked with bitumen! Light-colored woods such as maple are popular in narrow strip form and provide a perfect setting for modern rugs.

On solid floors there has been a long tradition of laying wooden blocks in various geometric patterns (parquet floor). Wood is softer and warmer underfoot than concrete and quieter on the ear. Today's range of hardwoods has extended the colors and textures available, and units of several blocks or mosaic panels are sold ready-mounted on fiber board and clear vinyl finished. In some cases a layer of cork is incorporated beneath the wood to increase resilience and insulation.

The warm, golden-brown colors of natural wood represent, for many, the most attractive feature of this type of floor. It is worth remembering that wood in strong light mellows in color with time. But more colorful wooden floors are possible using colored stains, sometimes incorporated in a varnish. These range from the usual wood colors to vibrant reds and greens which can look very good in both traditional and modern surroundings. Wood is perfectly happy outdoors too and for steps and raised stages is ideal. Modern microporous varnishes are a great innovation – they weatherproof the wood but still allow it to "breathe."

The use of paint on floors has a long history, particularly when applied through stencils. In this way borders of decorative motifs can add an extra dimension to a plain floor and work well even on old boards. Stenciling and painted decoration on floors needs to be protected by several coats of varnish. Even the legs of furniture and particularly stiletto heels can wreak havoc if the floor is of softwood.

Plywood, hardboard and chipboard are all wood-based manufactured materials which are invaluable in their large sheet form as subfloors for wall-to-wall carpets and vinyl flooring. Used with flair and confidence they can become floor finishes in their own right. Plywood

Below left These oak strips, laid end to end on the long axis of the floor, are manufactured with a cork backing, making the floor more resilient and comfortable to walk on. The underlayer of cork also helps to reduce noise and increase insulation. The strips are prefinished with a non-slip PVC coating which also makes them easy to clean.

Below Straightforward new floor boards can be transformed with the thoughtful addition of paint stenciled decoration. The pattern of the boards becomes irrelevant when simple designs are repeated and the stenciled border tailors these to the room.

Left Stencils can produce naturalistic designs as well as formal ones and may involve a considerable number of colors. Varying depth of colors can be achieved by applying the paint with a spray.

Below left Even worn, boarded floors can be rejuvenated with stenciled patterns, here by bold, repeated motifs in contrasting colors.

Below Cork tiles make a practical kitchen floor covering. They are easy on the eye, quiet on the ear and comfortable underfoot. The natural color variation in cork avoids the monotony of a plain floor. Once properly sealed (often done during manufacture) cork is easily cleaned.

comes with natural wood facings of different colors and textures, and when these are cut into geometric shapes and juxtaposed, the look of an expensive wood mosaic floor is recreated at a fraction of the price. Chipboard, too, can make inexpensive parquet flooring, although if water is spilt, it soaks up easily at the edges and even when dried out will be rough and weakened.

Cork This is a perplexing material – it keeps the sparkle in a bottle of champagne and gives a real bounce to the bathroom floor! Yet many people are unaware of its origins as the bark of a tree. The tree is the cork oak *Quercus suber* which thrives as a native in the sunny climate of Spain, Portugal and North Africa and is cultivated in the western United States and in India. The bark, or cork, can be peeled off and new bark grows in its place. Every nine or ten years for 200 years it is possible to crop this most accommodating and renewable of natural resources!

Cork as a floor material originated in Sweden when the by-product formed from the waste left from making corks for bottles was put to this use. Initially cork mats and later cork flooring tiles were made from the ground

cork glued together. Today the cork is often pre-finished with a protective layer of strong, clear PVC.

The color range of cork is now extended from dark and light shades of brown to grays and beiges. The warm tones and endless natural variation of pattern make this a harmonious material and one that is easy to live with. But its most distinctive quality is its resilience and warmth. For just the same reason that you can't push a champagne cork back into the bottle, so cork floors spring back into place when shoe heels and the legs of furniture are moved.

CARPETS AND RUGS

Throughout history people have chosen to put something warm and soft underfoot. Initially this would have taken the form of cured animal skins, varying from thick-piled polar bear for arctic Eskimos to meaner goat and sheep skins for Tibetan tribes.

Once the technique of weaving had been mastered, there was no end to the phenomenal wealth of color, texture and pattern that could be married together in the name of a carpet. Even today many of the carpets that come onto the market are still being made using exactly the same methods as our oriental ancestors practiced thousands of years ago. The remarkable persistence of the craft is not surprising, as the carpet holds a unique place in the domestic life of the Orient. It is far more than a floor covering in the Western sense and may replace a great variety of items, from tables and chairs to cradles and shrouds.

The term "carpet" is almost as confusing as some of the intricate designs which are found on it. Generally speaking, in Britain and the United States, "carpet" refers to machine-made wall-to-wall carpeting and in Britain, to any individual piece which is larger than 9 × 6ft. Anything smaller than this is known as a "rug." There are two exceptions to this rule. A long narrow piece is a "runner," and a carpet measuring 12 × 6½ft or larger, but in that proportion, is a "kelleye." In America, however, the term "scatter rug" is used for small pieces and "rug" for large ones.

Although most floor coverings are expensive, carpet is one of the few materials where the opportunity exists to gather up the chosen investment and take it, nomadic

Above Typical designs for dhurries or kelims.

Below A colorful cotton dhurry can be used almost anywhere in the house.

style, to a new abode. The choice available to today's buyer is quite bewildering in terms of plain or patterned, loose or wall-to-wall, knotted or woven, tufted or non-woven and, perhaps most important of all, which of the ever-growing array of fibers is going to suit the requirements.

The foundation of all traditionally made carpets and rugs consists of warp and weft threads, the former extending from end to end of the carpet and eventually, when cut, forming the fringes. During weaving the warp yarn is kept under even tension on a loom while the weft, which binds the warps together, is passed from side to side as if threaded over and under alternate warps. This results in a flat-woven weft-faced fabric resembling a European tapestry. The cotton Indian dhurry is made in this way and is available in a huge range of colors, frequently exhibiting strong geometric designs, simple stripes or borders. Generally these are fairly cheap, though more expensive types are available which may include wool or silk in their construction. Because these rugs have no pile they can be used upside-down and make useful outdoor rugs.

The kelim (or kilim) also follows this technique, though it is usually woven entirely from wool. Kelims can be very beautiful indeed, with strident colors and naïve motifs and at other times, to suit a late twentieth-century palette, they can be found arrayed in the most delicate

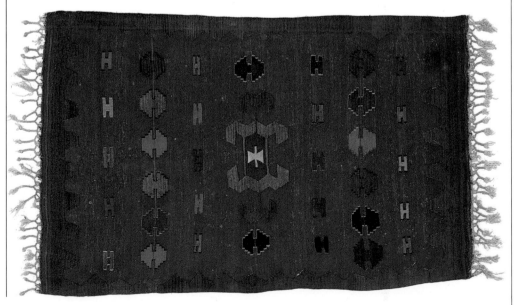

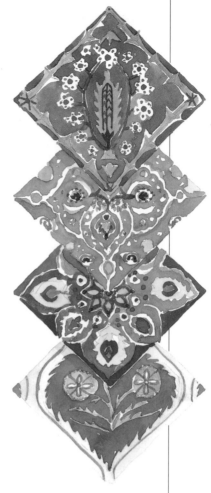

combinations of pastel colors. These latter kelims look particularly handsome on light-colored boards in a modern setting.

Pile rugs give a warm and cozy feeling to a room, providing something to sink one's toes into. They are made on the same principle as traditional carpets but the pile is introduced by so-called knots of yarn, usually wool, which lie between the foundation wefts. They are normally tied to every pair of warp threads and when the row is completed, the weft or wefts are passed through. All of these processes may be carried out by hand or machine. They may use natural vegetable dyes to color the yarn or modern chemical dyes which are far more resistant to strong sunlight, though it has to be said they will never fade to give the subtle luminosity of yarns dyed with vegetable dyes. The finest pile rugs are hand-woven and come from the Orient. Their designs are often copied for centuries on end and frequently bear highly symbolic motifs such as the dense floral designs which reflect the Persian preoccupation with gardens and the pleasures of Paradise.

There is another form of handmade pile rug which relies not on a warp and weft foundation but a pre-manufactured mesh base through which lengths of wool or other material are hooked and knotted. These are often made at home, sometimes by a whole family and, depending on the artistry of the "designer," they can be very elaborate affairs. Many fine examples, now precious heirlooms, were made in Britain and the United States, particularly at the beginning of this century.

Fitted carpets The range and diversity of rugs knows no bounds but sometimes a unifying factor is called for within a decorative scheme and here a wall-to-wall carpet really comes into its own. A smooth, plain carpet is often the most successful way of setting off furniture, pictures and *objets d'art* (and even rugs) without creating artistic mayhem in the home.

Carpets of this type are manufactured all of a piece; that is to say the backing is produced at the same time as

Above Traditional motifs taken from oriental carpets. The top example shows a stylized vase design, the middle two are medallions and the fourth (bottom) is a Persian flowers motif.

Above left Rugs are the most transportable of floor coverings and can be moved around with our homes and within our homes.

Top A light background and soft colors enhance the delicate motifs of this modern kelim. Because these rugs have no pile they are equally suitable for use on top of wall-to-wall carpets as well as wood boards.

Above right Wall-to-wall carpet offers a draft-free, unified floor covering.

Above A traditional motif from an oriental carpet. The subtle differences of color in the background adds to the richness of the carpet.

Above left Traditional oriental pile rugs displaying a wealth of colors; the subtlety of the rich colors relies on the use of vegetable dyes.

Left "Rug of many bosoms" by Jasper Morrison: the restrained palette of black and white sets the tone for this strikingly modern creation.

the pile. They may be woven as in the case of "Wiltons" or "Axminsters," or tufted as are the vast majority of patterned carpets in the United States. In addition there are non-woven carpets where the fibers are held in place with adhesive on a pre-manufactured backing.

The key to quality and durability of all these carpets lies largely in the amount and type of fiber used per unit area. There are many fibers and combinations to choose from. They include 100% wool, 100% synthetic or a mixture of the two in varying proportions. Wool is generally the most expensive but compensates by demonstrating excellent characteristics in terms of appearance, durability and flame resistance. Wool will also happily combine with other fibers, including synthetics, to produce some very hard-wearing and proportionately cheaper carpets. The main disadvantage with synthetic fibers is their low resistance to flame, and they never look quite as refreshingly buoyant as wool. There should be no difficulty in identifying either the composition or the type of construction of any modern carpet, as they now all carry that information on a label that is usually attached to the backing.

For the ultimate in flexibility, carpet tiles provide a satisfactory alternative. They are available in a wide range of fibers including wool and other animal hair and are backed with a flexible material such as PVC, rubber or bitumen-impregnated felt. Laid in a single color, these tiles will resemble broadloom carpet but endless arrangements can be adopted using two or more colors. For areas of hard wear or access to under-floor services these tiles, if loose laid, are ideal and can be moved around to distribute wear.

Caring for carpets All carpets, except carpet tiles and foam-backed synthetics, benefit from a good underlay. It will not only make the carpet feel better underfoot but it should also prolong its life. Underlay also helps to protect the carpet from the back by cushioning it against uneven floorboards which could soon abrade the backing, and underlay cuts down the amount of dirt which works up through gaps in floorboards. It also helps heat insulation.

The life of all carpets will be enhanced by basic good housekeeping; they should be cleaned regularly to remove abrasive grit and dirt. Wall-to-wall carpets can only be vacuum cleaned or brushed, but loose rugs benefit from shaking outdoors. Most carpets can be gently shampooed, but check the manufacturer's instructions if possible and always test a small area for color fastness and shrinkage first. Never allow the carpet to become saturated and always try to deal with spills and stains as they occur.

Matting Carpet doesn't only come from the animal kingdom. Matting of various sorts provides a vegetarian alternative if you are looking for cheap yet effective floor covering. Sisal, coir, rush, seagrass and maize are all natural plant materials; some have a tendency to shred.

Undoubtedly sisal and coir are extremely hardwearing. Sisal comes from the leaves of the agave plant and coir from the outer husks of coconuts, and both are relatively cheap. In their natural state colors range from pale to dark brown, but sisal is often dyed in an attractive range of colors and patterns. One problem that befalls all types of matting is the amount of dirt and dust which inevitably falls through to the floor beneath. If the area of matting is small enough to pick up and clean underneath, then there is really no great problem, but in larger areas

this can be a drawback. Better-quality forms of matting have a latex or vinyl backing, allowing the removal of dirt particles with a vacuum cleaner.

Woven seagrass is an attractive form of matting and its knobbly texture responds well to imaginative lighting. Like any natural matting it is not particularly kind to bare feet but it does alleviate the icy coldness of a ceramic-tiled floor. Rush and maize mats usually consist of patterned squares of the material sewn together with twine to form mats. All forms of matting look particularly good in areas such as conservatories or sun lounges where their color and material complement the outdoor world.

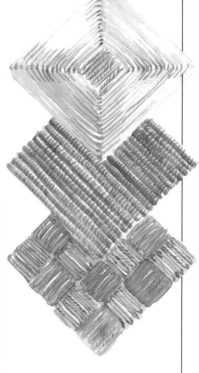

Above Matting is made from various materials providing a choice of colors, textures and patterns. The examples shown here are: a seagrass square (top), heavy-ribbed coir (center) and maize matting (bottom).

Top left Strips of natural sisal matting from a roll, laid and stitched together, provide the perfect background for this pretty bedroom. Sisal, like other fiber matting, can be hard on bare feet, but the soft cotton rug between the beds offers a more comfortable stepping-off point.

Left Practical and hardwearing ribbed coir matting tiles are an economical solution for this modern dining area.

Chapter Three

HALLS AND STAIRCASES

MAKING AN ENTRANCE

O ften dark, difficult and dull, the hallway provides a visitor's first impression and accordingly should be dressed to impress. Since the hall tends to be virtually all floor, this is going to be the most important decorative detail. If it is dwarfed by a tall ceiling, the floor needs to be given strong, eye-catching treatment. Hence the stark black and white tessellated tiles of the nineteenth-century entrance hall, or the kind of gothic sweeping staircase on which Scarlett O'Hara made her grand entrance.

Staircases have been used to create the focal point – the grand feature of the main hall – ever since they became a piece of decorative furniture and not simply a ladder to the upper storeys. The grander the house, the more spectacular the staircase, from the giant carved and balconied double staircase of the Elizabethan manor to the more modest but no less ornamental curly features of the Victorian townhouse. They perhaps reached their decorative peak in the eighteenth century when the wealthy property owners of Britain and America com-

missioned the best architects of the day to design their homes and every feature in them in total Palladian splendor. From turned and carved wood to decorative cast iron, it is only recently that the stairs were not positioned and designed for maximum impact.

In order not to ruin the effect, stairs and hallway floors should be planned together. That is not to say they should always be given the same treatment, for this is not always practical or desirable. Floor covering for steps should be non-slippery and securely fixed, yet perfectly in tune with the general style of the staircase. Providing the eye is led smoothly along the hallway and up the sweep of the stairs, it doesn't matter whether the hall is sensibly clad in coir with the occasional exotic rug, while the stairs are polished wood with a central stair carpet, so long as the chosen colors and tones make that visual connection. Equally important, maybe more so, is the process in reverse. Many hallways are designed to be viewed from above, within the framework or curve of the staircase, and designers, both past and present, have

exploited this advantage to the full with a stunning detailed design in carpet, marble or ceramics which follows the shape of the hall. An aerial view of such a design really shows it off at its best.

Still viewing the situation from top to bottom, landing floors can be chosen with the same parameters in mind as for hallways, with a slight adjustment in practical requirements. The hall must be designed to take a great deal of wear and tear, not just from one room to another but from outside, so any ground surface should be tough and easy to clean as well as stylish. Upstairs on the landing, you will be looking for comfort, a soft transition between bedroom and bathroom, something that can take a reasonable amount of traffic but remains looking luxurious. Sandwiched between, the stairs offer a limited choice of practical options, mainly for reasons of safety. Marble or ceramic tiles, for example, may be fine on a couple of short, shallow steps, but a complete staircase, especially a steep one, would be lethal in a family home. The wall-to-wall carpet, continued from the hallway, up the stairs and along the landing is an all too-popular solution, but unless you commission tailor-made borders and patterns to add a little oomph to the scheme, rather a dull one. Classical or modern all-over designs can be used to add more interest for those who don't have the budget for custom-made carpeting; while a strong color or a small pattern or fleck wall-to-wall can be useful in small areas to maximize visual space. A patterned rug, square, rectangular or circular, depending on the shape of the hall (remember that aerial view), is useful for adding impact in selected areas and might be in a rich oriental style, understated modern pastels or stark primary geometric, according to personal preference and the style of the staircase.

Often a classic staircase is better suited to a stair carpet, a narrow strip of plain or patterned carpet which runs up the center of the stairs, the treads stained and

Right Hard-wearing but highly attractive, these lovely antique tiles are reminiscent of a French farmhouse.

Left A stenciled border transforms plain polished boards. Subtle color schemes enhance the effect and bring out the grain of the wood.

varnished or painted on either side. The carpet is usually held in place with gripper rods for safety and the "framed" effect this produces is attractive, especially for oriental runners. Similar rugs and runners could be used in the hall and on landings, anchored firmly in place with padded, adhesive underlays, on stripped and polished boards, cork tiles or new wood flooring. This latter material makes a fine hall floor, dark polished oak parquet for a traditional home perhaps, or bleached pine or sleek beech strips or blocks for a smart contemporary interior. Cork is less sophisticated but it is warm and comfortable, quieter than wood underfoot – excellent for a busy family home. Both are highly practical and easy to keep clean.

If you are lucky enough to live in a house with an original old staircase in good condition, you may prefer to leave stone or wood treads unadorned where histori- cally accurate. Never mind the cold feet and clatter, the staircase will be displayed in its proper glory and you will probably be keen to keep the hallway similarly authentic. Replace or restore stone flags or polished wood as appropriate, adding rugs or rush mats. Other decoration or furnishings will be minimal; the very simplicity of the flooring sets the atmosphere.

CREATING AN ENVIRONMENT

Where the staircase isn't characterful enough to suggest a suitable treatment, or when you need further inspiration to choose a flooring style for hall and stairs, you should consider the kind of atmosphere you are trying to achieve. This isn't an area to spend a lot of time in, but it is a place to receive callers, a thoroughfare linking all parts of the house, and the way the area reflects your lifestyle is important. It could be a relaxed and friendly place, with comfortable chairs, prints on the walls and an umbrella stand. Or maybe you prefer a smart, slick, almost anonymous environment: everything in its place and the rest of the rooms private behind closed doors.

Should you be dissatisfied with its shape and size, it is possible to play visual tricks in a hallway by laying flooring with a strong directional pattern across, rather than along a passageway to create the impression of extra width. Stripes and squares, or even plain tiles set across the diagonal also serve to confuse the eye and disguise an area's true shape. Strong dark colors and busy patterns will bring walls together and lower high ceilings; an all-over light, plain or small-patterned floor covering can make the area look rather bigger than it really is.

This may not be a part of the house in which you linger, but the floor will be taking considerable wear and tear, so however stylish the chosen surface hopes to be, it must also be tough and easy to clean, safe and non-slip. Luckily the options are wide enough to suit all tastes.

We have already looked at the classic combination of polished wood and rugs. Equally traditional for hallways are tiles, marble and stone slabs – expensive but long-lasting, with a better choice of color, texture and pattern

Below Sisal can be fitted wall to wall like carpet to provide a natural, all-over flooring that is tough and textural – a good neutral background for stronger elements.

than ever before. This is the perfect place to plan a pattern of shaped French, Italian, or Mexican terracotta tiles with contrasting corner inserts of black, red, green or tiny-patterned ceramic squares for a sophisticated farmhouse atmosphere, or opt for nineteenth-century elegance with an arrangement of black and white tessellated ceramic tiles. Marble is another classic: difficult to lay and slippery but now available in tile form. These are the materials to choose in country homes where mud is a constant visitor, or for the beach house besieged by gritty sand. In hot climates the coolness of stone or fired clay underfoot will be welcome; soften and warm the floor

Left Popular with early American settlers, painted floor cloths like this fine example are surprisingly hard wearing and were often seen in halls and passageways where they added color and comfort to rough stone floors. This plain piece of canvas is richly painted in imitation of an oriental rug design.

Right Hand-painting adds an extra perspective to any material: the ceramic tiles on this long and narrow passageway floor have a softness and mellowness that make them look as if they've been there for centuries.

with underfloor heating or thick rugs in winter. No other flooring can offer such toughness and easy-clean qualities combined with unlimited scope for colored pictures and patterns, borders, panels and designs, from tiny terrazzo and mosaics to the large checkerboard effect of the classic tiled hallway.

For homes with less stringent needs, there is the alternative of fake marble, stone, cork, tiled and wooden flooring in the form of vinyl tiles or sheeting. The more you pay, the more realistic the effect, with the option of choosing ready-made popular designs (terracotta tiles, black and white checks, Mediterranean ceramics, mosaic, parquet) or putting together your own ideas from component parts. Soft, warm and easier to lay than the real thing, vinyl and its forerunner linoleum – which is coming back into fashion, particularly among those who like to devise their own designs as it is easily cut and fitted into new shapes – both make excellent choices for stylish hallways and passages.

ALTERNATIVE CHOICES

There are smart alternatives to rugs and carpet too: coir matting can be laid wall to wall for a rough, tough country-style atmosphere. Natural shades are pleasantly neutral, a good background for exotic rugs, or choose one of the more decorative weaves and earthy color combinations of reds and greens. The only drawback with coir and other grass type mattings is that they shouldn't really be continued up the stairs as they become slippery at the edges with use. Another impressive nineteenth-century idea worth copying is the floor cloth, popular in Britain and America, particularly in Pennsylvania where some fine examples are still to be seen. It is a cheap alternative to more expensive rugs, tiles and carpets. An appropriate design is copied onto a piece of canvas with paint, heavily varnished and fixed to the floor over a protective pad to soften and prevent uneven wear. The floor cloth is surprisingly hard-wearing and remarkably effective. It used to be a common choice for smart hallways and probably developed in time into the backed and printed linoleum of the early twentieth century: interestingly, it is coming back into its own as an original and creative form of floor covering.

Right Oriental runners are specially designed for long narrow halls with tall ceilings. They are shown off best against a plain natural flooring like these stripped and polished boards. The strong color and pattern of the rugs is sufficiently attention-grabbing to balance the somewhat lofty proportions of the hall.

Above Clever co-ordinators should look out for complementary patterns and designs like these. Full use has been made of borders and cut-in shapes to create a stylish mock marble hallway in vinyl. The theme has been continued up the stairs in comfortable, quiet carpet. It is a treatment that enjoys the best of both worlds without reducing style.

Right The classic entrance hall consists of a vast expanse of floor and large double doors opening onto elegant reception rooms. It cries out for these marble tiles with contrasting cut corners or the best vinyl tile equivalent. Where the area is too extensive for strong black and white contrasts and converging parallel lines, go for more subtle gray and break up the area with a detail of special interest in the center like this simple inset.

Above Landings take less wear and tear than hallways and can afford a more luxurious treatment for bare feet in the morning. Wall-to-wall neutrals like this carpet are guaranteed to blend perfectly, not just with the immediately surrounding decorations, but also with bedroom furnishings in rooms that open onto the area.

Left A grand entrance needs a stunning floor, tough enough to take the wear and tear of centuries. There's no mistaking the real thing with this splendid irregular checkerboard of polished marble.

Left Don't forget that entrances are often viewed from above as well as below. This understated all-white marble treatment shows how stunning it really is when seen from the stairs or landing. Restrained use of other furnishings and accessories – simple green foliage plants against the white – reinforces the impact of the geometric design.

Above left Wood is a superb, adaptable flooring for any part of the house. It provides the perfect practical option for living room, hallway and stairs.

Above Wood flooring makes the transition between house and garden with only a sheet of plate glass between a wood decking and polished planks. It can be inexpensive – strip and stain existing boards or give a real luxury treatment by choosing a new floor in a special wood such as oak or beech. Both these homes have made full use of rugs in a variety of styles to soften the surface and add color and pattern.

Left Mixing hard and soft surfaces adds interest in a large hallway or entrance and helps to identify different areas where the space has to serve several purposes. Here the carpet sweeps down the stairs and is neatly continued around a small welcoming area, before meeting practical ceramics closer to the entrance doors. The result is far more attractive than if the stair carpet stopped abruptly at the bottom of the stairs.

Above Sometimes it is enough just to let shapes and materials speak for themselves. Sturdy ceramic and russet terracotta tile passageways offer the perfect contrast to an architectural arrangement of plain painted walls and supports and deep carpet in flat colors.

Below Bleached or stained and polished wood makes an elegant hard-wearing hall floor. In this airy entrance, it has an extra relevance, continuing the theme of wood paneling on the walls and ceiling. It is also highlighted by a checkerboard pattern at the door. This area is large enough to take this all-over wood treatment without looking like a cabin.

Left Make the most of a grand entrance by covering the stairs in a bordered stair carpet. This intricate design imitates the curly wrought iron work of the stairs themselves. The colors of the carpet have been picked out for paintwork and in the vinyl floor of the entrance hall to maximize the effect.

Above Some staircases are stunning enough to need no adornment except for the warmth and color of a piece of fabric and mellow furniture, like this stone and metal construction in an old farmhouse.

Right Carpet borders are useful for highlighting stairs and landings. Here a strong wide design adds weight and interest to the all-over pattern.

Below Well-worn brick is the perfect foil for natural wood and stone. It can be softened with inexpensive rag rugs which are easy to take up and clean.

Right A large hall with a dramatic sweeping staircase can take the richness of an oriental carpet. They look their best on a natural wood floor, stained and polished to equal richness.

Right Different colored woods have been used to create an elaborate border in parquet that is both subtle and stunning. Although decorative, this is a floor that will wear well and always remain looking good.

Left The inspiration here is classical and timeless, but the easy-care vinyl tiles give this entrance a spacious, contemporary feeling, a perfect foil for the richness of the walls and furnishings.

Below In spite of the opulent patterns on almost every surface of this William Morris interior, this landing carpet holds its own, its intricate design echoed in the windows at the end of the hall.

Right The owners of this small town house have chosen a velvet pile carpet for the stairs which makes a strong textural contrast with the sisal matting downstairs. The treatment works well because the shades of beige are offset by the cream and white of the fireplace and bannister. Notice how the staircarpet has been folded at the junction of each riser and tread for a perfect fit.

Chapter Four

LIVING ROOMS

◇

LIVING IN STYLE

Once the drawing room was the decorative climax of the house. It had the grandest moldings, the best furniture and the most splendid floor: the ideal setting in which to sit in splendor or dazzle your guests. The living room remains the part of the house where we are prepared to spend most money, to represent the main reflection of our taste and style, and where we agonize most over the furnishings. It is easy to forget that the floor constitutes one of the largest decorative elements in the room and that the wrong choice could ruin the effect of the whole scheme. There is a temptation to skimp on cost – to forget the tough treatment we expect the floor to take and the length of time it will reasonably last. Poor quality will quickly look tired and need replacing; an unsuitable choice may be difficult to live with over the coming years and date too quickly.

It is worth spending the time to choose wisely, making the floor the starting point of your scheme, deciding on the best material, color, pattern and texture for the room and then selecting other furnishings to suit according to the most effective contrasts and harmonies of color and texture. Flooring as an afterthought is destined for disaster.

Take a look around some of the loveliest old houses and you will find that as much care and attention has been afforded the floor as the grandest architectural features. An oriental rug may be specially woven; a mosaic or marquetry floor specifically designed to mirror the shape and pattern of moldings on the ceiling. These days we are not always allowed the indulgence of a living room exclusively set aside to show off our prowess at interior design. The room may double as family room, playroom or dining room, and the decorations, particularly the floor, must be selected to satisfy more than one criterion.

If you need a tough, easily swept floor that is multi-functional yet still attractive, or if you are expecting to move within a short space of time and are reluctant to spend money on a flooring you can't take with you, a polished wooden or parquet floor with a

Previous page This carpet with a stylized pattern of ribbons, garlands and swags, perfectly complements the formally draped curtains and the border of the carpet emphasizes the room's spaciousness.

Below This rug laid over white marble adds color and warmth and helps create a focal point for the seating area. The pattern of the rug echoes the shapes of the tiles, the cushions and the table.

selection of rugs is the best option. Similarly, a wall-to-wall velvet carpet, so easily marked and soiled, is not for homes with young children and pets. It must be practical but it must feel comfortable too. Your lifestyle will dictate this: while a sub-tropical or Mediterranean climate cries out for cool marble, terracotta or ceramic tiles, the stone floors of a Scottish castle need the thickest rugs. A musician will look for a flooring offering the best acoustics; bamboo matting is for those trying to reproduce the calm serenity of oriental style.

In living rooms serving more than one function, a mixture of flooring materials can sometimes help delineate the different areas. This works best where the contrast isn't too striking – a hard surface butted onto a soft one for example, but sharing the same shade, pattern or design. Care should be taken when visually dividing a small room; it may be preferable to use one type of flooring throughout to create the impression of space and rely on a variety of lighting effects to change the emphasis according to the room's use. Other canny space-saving ideas include running hard surfaces such as cork or wood up the wall at skirting level or continuing wall-to-wall carpet over built-in bench seating to blur the true boundaries of the room.

FINANCIAL CONSIDERATIONS

Lack of funds is no excuse for boring flooring in the living room; some of the most exciting ideas are born out of the necessity to come up with creative designs using the materials available. Providing the floorboards are in good condition, they can be stripped of dirt and marks using a rented machine and stained, polished or painted in an infinite variety of styles. Older properties may still have the original oak, elm or pine boards and these are worth waxing and buffing up to become a feature in themselves. Nondescript pine or oak can be stained or bleached in subtle pastel shades – the more coats of stain, the stronger the color – or painted a bright glossy primary shade. Stencils add further decoration, excellent for borders around the edge of the room or to create panels and patterns. Design and paint your own floor cloth, fasten it to the boards and stencil a matching design on the flooring for the border. Top with several coats of varnish for durability.

Ground-floor living rooms do not always sport a suspended wooden floor, but a sound, level concrete screed is equally suitable for creative paint treatment. Work out a strong geometric design, or go for an elaborate *trompe l'œil* complete with mock oriental carpet and fake "wooden" borders. This would be the perfect opportunity to experiment with some of the broken paint effects such as splashing, stippling, flecking and sponging; a hand-marbled floor would look magnificent at a fraction of the cost and upheaval of the real thing.

Other less expensive options include cork tiles. They are soft, easy to lay and perfect for family living rooms; a few rugs add a more sophisticated air. Cork needn't be a brash, cheerful honey glow these days either. Color options include subtle grays and beiges which can look

stylish. You can always stencil designs on the tiles and give them a couple more coats of varnish as described for bare boards. In fact if your boards aren't in good enough condition for display, cork makes an excellent alternative flooring. Natural coir matting is equally useful over large areas as an all-over, subtle background for rugs or mats. It is tough enough to withstand the rough and tumble of country living or a rigorous household and has an interesting texture that might appeal to those tired of shiny smooth surfaces or the stifling pile of carpeted floors. Coir is available in special decorative weaves and patterned color combinations, but this is more costly.

A step up from polished boards is a new wood floor, arranged in strips, planks, blocks or parquet patterns in a wide choice of woods and finishes. This kind of floor

Below In a basically all-white room the main function of the kelim is to provide color, but its pale shades are subtle enough not to disturb the tranquil atmosphere.

always looks classy; it is a little noisy underfoot but easy to keep clean and, should you tire of it, disappears gracefully under rugs or carpet. The excellent variety of woods available means you can adapt the look to whatever style you desire, from a classical mellow glow to streamlined Scandinavian-style simplicity. An eye-catching result relies more on making the right choice than coming up with all the creative ideas and doing the hard work yourself.

THE POPULAR CHOICE

Old-fashioned flags or terracotta tiles are only for living rooms in rigorously authentic period properties. The most popular choice of flooring for living rooms must be carpet, a term that covers an almost infinite choice of options. It's not only color and pattern that has to be decided, but style and texture, length of pile and the possibilities of special designs. There is endless scope for harmonizing and contrasting colors and textures with other furnishings, from the standard short-pile carpet (unobtrusive and hard-wearing over large areas) to luxury velvets and rather impractical shag pile that is so long and thick it needs a special rake to groom it. The way the pile is cut determines its texture, and you will have to balance your choice against shiny chintz, rough tweed or textured jacquards, glossy paint, soft wallpaper and other elements of the room design. Almost any color is available, from vivid primaries and popular fashion colors (take care to select one you will be happy to live with long after it has gone out of vogue) to the natural wool shades which have a pleasing fleck that remains reasonably anonymous yet doesn't show the dirt.

Near-neutral carpeting is all very well in living rooms where you want to make a strong statement with special rugs or designer furniture, but more and more people are becoming adventurous enough to experiment with the new patterns and colors (imitating the best modern wallpaper designs) and the option of adding borders and inset designs to suit the shape of the room; a return to the principle of the elegantly designed drawing room. If you have the patience, there's nothing to stop you cutting and creating your own carpet patterns; the spectrum of colors available is certainly sufficient to give enough

Left Here close-textured rugs provide a counterpoint to glass, chrome and ceramic tiles, while their deep colors and severe geometric shapes are perfectly in keeping with the room's style.

choice. Alternatively, let your imagination loose with several shades of carpet tiles, clipping the corners and adding insets of a contrasting color in imitation of ceramic tiles, or cutting the tiles diagonally into halves or quarters to make up stunning geometric patterns.

One advantage of a more neutral background flooring is that you can go to town with rugs and make changes at a whim. Here again the scope is tremendous, from the best richly patterned oriental rugs – the largest can replace a wall-to-wall carpet, although they look best with a border of stone or wood floor to set them off – to the flatter, woven dhurries and kelims, shaggy rag rugs in cloth or leather, thick sculpted plain wool rugs based on the classic designs of Fair Isle sweaters, goats hair and sheep's wool, thin cotton and woven grass or bamboo, or, for a piece of flooring art, a one-of-a-kind designer rug. Replace them, swap them, trade them in; hang them on the wall when you get tired of them – rugs are probably the most exciting flooring element, appealing to the nomad in us all.

Below A floor of tiny ceramic tiles would not often be considered for a living room. But its light-reflecting qualities are put to good use where there is sunshine and large windows.

Above Stone tiles produce sufficient natural variations of color and tone to maintain interest in a room dominated by bare brick walls and a strong assortment of traditional furniture. Underfloor heating ensures the stone remains warm to the touch.

Above Polished wood boards maximize the length of a room and provide a practical multi-purpose surface. This living room opens on to the garden, requiring a floor that is easy to sweep with rugs to add warmth and softness.

Right Brighten up a plain carpet with a few carefully chosen rugs. This bold fireside mat balances a tiny ornamental metal grate and old-fashioned painted furniture.

Above left Mixing hard and soft surfaces works particularly well in hard-working living areas like this garden room. Terracotta tiles link inside and out and a large pastel dhurry adds comfort to the seating area.

Far left Sleek ceramics run right through the ground floor of this modern home; their creamy stone color reflects light and provides an excellent soft-toned background.

Near left Patterned carpet can be dominant, but the subtle shade variations in this soft pink carpeting are just enough to add interest over a large area.

Above The natural warmth and grain of the floor and ceiling, furniture, doors and windows needs little embellishment save an excellent leafy view and the occasional splash of color.

Left Some floors are deliberately designed to add a dramatic touch to an interior. A totally monochrome theme may seem too stark, but the large black and white tiles throughout the ground floor are busy without being overpowering. By laying them against the direction of the rooms, they seem to open out the area and avoid any dull parallels.

Below The stylish wood floor adds color to a simple decorative scheme. The wooden furniture emphasizes the honey-colored tones.

Right Natural wood needn't be treated to a warm honey glow; in this pastel shaded living room, the narrowness of the boards and paleness of the wood is exactly right as a background for an individualistic collection of furniture and furnishings.

Below Start the interior design with the floor, and you can build up an interesting blend of textures and finishes in the living room. Slate tiles sit happily with natural wood and a pair of large soft sofas.

Above Rugs aren't simply useful for softening a hard floor surface, they can be a work of art on the wall as well as the floor. This bold blue and white treatment is instantly eye-catching.

Left A specially designed carpet border, following the contours of the room along skirting boards, into bays and around fireplaces, will make the most of elegant architecture. Designs are usually made up to order and set into a plain carpet. Here a wide border mimics a deep skirting board and moldings around the room.

Below left The focal point of the room is a magnificent molded ceiling, and other features have been played down to give it proper prominence. The plain creamy carpet balances the cream painted ceiling, and incorporates a sculptural pattern that resembles a stone tile floor. The design is sufficiently interesting without being dominant.

Right Wall-to-wall carpet in a neutral shade is a safe, long-term choice in a living room but can look rather dull even when fabrics in the room are strongly patterned. A large rug highlights the hearth area and provides a natural focus around which to arrange comfortable seating furniture. It also adds new patterns, colors and textures to the total arrangement.

Left Marble may seem a strange choice for a living room, but with a large rug and the right furniture it can look stylish and you can make changes easily with a simple change of props.

Below The traditional black and white marble floor looks stunning in a classic interior, particularly when laid more interestingly on the diagonal across the floor space. This quality vinyl mock marble looks equally stunning and is a lot more comfortable.

Right Rugs are frequently used to define the hearth-side area in the living room, especially where the room is large and multi-functional. Here a wood floor makes sound sense for a room with a grand piano, but the area around the fire is given a cozy, more comfortable feel with a large pastel blue carpet.

Below Parquet is another classic flooring that looks good and lasts for years. The wood is laid in patterns to produce a decorative effect. This brings out the tone and grain differences in the wood which sits well with upholstered furniture and traditional furnishings.

Stenciling is a wonderful technique for floors – the perfect opportunity to experiment with panels, borders and dramatic centralized designs. If you are keen to try it yourself, the only rules are to plan your design carefully before you start so that it fits the room and its features, and to protect the design with plenty of coats of varnish.

Soft, muted or natural colors seem to work best on stripped boards, hardboard or even a new wood floor: in blocks and stripes (left), or the dramatic tattoo (below).

Left and inset Soft shaded borders can be used to trace the limits of a room such as the threshold. They can take on an almost three-dimensional effect if you use spray paint.

Chapter Five

D I N I N G   R O O M S

THE GOOD THINGS IN LIFE

The dining room always used to be a separate room for entertaining or having a formal dinner with the family. More recently, the room has become something of a chameleon, changing its purpose according to the area available and frequently sharing floor space with the kitchen, family room or living room. This makes the choice of furnishings more difficult, especially floor surfaces, which may have to combine the tough practical properties required by a kitchen with the more decorative needs of a dining room, or be robust enough to withstand dining spills, yet not spoil the ambiance.

The dilemma can be solved in one of two ways, depending on the style and size of the room. One solution is to select a flooring that can be used throughout, something like wood, cork or ceramic floor tiles that look smart yet can take plenty of punishment and be dressed up with rugs in areas needing a softer touch. The alternative, really only practical in larger rooms which can take being broken into smaller areas, is to use a change of flooring for the separate functions. This can be effective with sensitive choice of materials and often a change of level will successfully emphasize the effect. The dining area may be raised on a small dais or platform covered in wooden strips, ceramic tiles, or one of the special vinyl effects; fake marble could look spectacular set into a black carpet as part of a dramatic living room.

A change of level and contrasting flooring serves to underline the difference between two areas serving two different functions, but where the room is small or you prefer to make the dividing line between dining and relaxing or food preparation a little less obvious, more harmonious companions might be chosen. For example, checked ceramics in the kitchen could give way to softer, matching checks in carpet tiles around the dining table. Clever paint effects in the immediate vicinity of the dining table, such as painting bare boards, or hardboard in imitation of the patterned carpet in the main part of the room, is an original idea. Alternatively, cork or patterned vinyl might be a close match for the carpet.

THE SEPARATE DINING ROOM

Those lucky enough to enjoy the luxury of a separate dining room have no need of such illusory tricks. Their sole concern is that the dining room floor be smart and comfortable, yet easy to clean. There is no point in fitting the room up with an expensive, thick-pile carpet if you have young children or spend dinner parties in an agony of expectation in case a guest knocks over a glass of red wine. There are more practical yet no less stylish options: carpet is a possibility, providing the pile is sensibly short and the fiber treated with one of the new stain-resistant systems. The alternative is to use carpet tiles, available in a far better choice of colors and styles these days and having the advantage that you can lift and replace any that have worn, or even run them under the tap should an accident occur. Different colors can be easily arranged to create simple patterns. More effective is to take advantage of one of the custom-made carpet options where you can combine various elements to make borders, panels and patterns. Use them around the edges or to set off the table in the center – an old fashioned but highly effective

device that adds a really classy touch to the dining room and makes the most of a fine dining suite. The technique is easily adapted to a more modern scheme by choosing bolder colors, stronger contrasts and geometric designs.

You could of course use the same idea in other materials. The budget dining room with stripped and varnished boards has no need of rugs or further decoration if there is a stenciled border outlining the table area on the floor and matching stencils on walls. Ceramic flooring tiles offer plenty of border and pattern ideas and always look smart and stylish, if rather chilly in the dining room. They give that unmistakable Mediterranean atmosphere and are naturally ultra-resistant to food spills. However, like tiles and stone flags which look superb in a farmhouse kitchen/dining room, dropped glass or china is guaranteed to smash on impact. Terracotta tiles and flagstones should be sealed against staining where food is to be prepared or eaten.

Non-purists will be happier opting for one of the more realistic vinyl options, available in tile or wide-sheet format for easy installation. They provide a warm, soft surface underfoot, yet possess all the visual impact of stone, marble, terracotta or ceramics. It is better to avoid the cheaper styles, as these may scuff through the constant movement of dining chairs. Equally practical and comfortable are cork tiles; easy to wipe down and easy to lay, especially into awkward shapes. They are now available in a wider range of colors and have an advantage in the dining room in that they spring back into shape when furniture is moved to a new position.

Cork may be quieter, but you can't beat a new wood floor for style in the dining room. Parquet, strip or block, bleached or buffed to a honey glow, wood makes an unexpectedly good foil for quality wooden dining furniture, is easily swept or wiped clean and lasts for years. It is an excellent companion for all kinds of rugs, providing they are properly secured to the floor to prevent them slipping – useful in dual-purpose living/dining rooms or in over-large dining rooms where a wide expanse of wood floor can look daunting.

Two very different ways to use rugs on wooden floors: a selection of small, subtly patterned Persian rugs and runners (above) break up an extended floor and indicate areas for dining and natural passageways. In the cozy dining room (right), a large, Chinese-style carpet in strong colors provides the inspiration for deep blue walls and touches of greenery. Even with large carpets, it is a good idea to leave a border of floor showing all the way around the edges to set off the design and color.

Below The dining area can be defined in many ways, although with a circular table this may be more difficult. This modern home solves the problem by sinking a small circular well and carpeting it – a comfortable contrast to the glossy slate floor tiles on the rest of the ground floor.

Below Tiles lend themselves to all kinds of design combinations which can be picked up and adapted to the other furnishings in a room. An all-over floor in warm red marble tiles with diamond inserts has been teamed with glossy red and black dining furniture to provide harmony between the new style and the old.

Right This pretty kitchen diner has a creamy color scheme continued across the floor with a subtle shaded pattern in a bordered carpet.

Left Terracotta tiles can provide unexpected scope for decorative effects if they are positioned slightly out of line to incorporate tiny colored ceramic inserts. This adds that extra touch in a room that doubles as kitchen and dining room and therefore needs something more than just a functional floor surface.

Above In a large kitchen you can sometimes get away with a handsome rug to delineate your dining area, providing you maintain an area of easy-clean terracotta tiles or similar around the work areas. The natural shade variations of terracotta are a good foil for subtle oriental patterns.

Right A plain wood floor is more than a match for both kitchen and dining areas, offering as it does a tough but warm surface. In this kitchen, the floor cleverly makes the link between a wooden table and chairs, and the soft, hand-painted kitchen units.

A conservatory or garden-room extension can be pressed into service as a dining room where a hard-wearing, easy-clean surface is essential. Terracotta and ceramic tiles are highly decorative, and they are easily swept and washed of garden and food debris. The patterned ceramics look perfect in a Mediterranean-style grotto (right); while rough, mellow terracotta tiles provide a change of texture and tone on a rather sophisticated terrace (below).

Left Cork flooring makes good sense in a family home. It is soft and warm, easy to lay and practical to clean. It always looks stylish and is an excellent companion for natural wood.

Below A light dining bay has been given a spacious feel with plain white ceramic floor tiles and the simplest of dining furniture and decorations. The tiles have been neatly butted onto the carpet which covers the main part of the room in a way that makes a feature of their shape and position.

Top left Tough practical flooring in a beach-side house is softened with a large, woven cotton rug which can be taken out and shaken as required. Its subtle shades cleverly pick up the colors of the seascape beyond.

Top right An elegant dining room with fine furniture can take the luxury treatment of a carpet. In a long narrow room like this one, a plain neutral shade was the obvious choice, making the handsome dining suite the focal point of the room.

Above Stripped and polished boards make a warm sensible flooring in the dining room – they seem to suit every style of home. It looks equally in keeping in the grand house with its tall windows, as in the first floor apartment furnished with a happy mixture of acquired furniture (right).

Right Traditionally, the grandest dining rooms would offer the chance to enjoy a leisurely meal with the very best oriental carpets underfoot. The classic border design is perfect for setting off the shape of the room with table and chairs in the center. In a large room like this one, exotic patterns and rich colors make a feature of the extensive floor area.

Chapter Six

K I T C H E N S

◆

EQUIPPING THE WORKPLACE

Contemporary kitchens have become so design orientated with fashion cabinetwork, co-ordinated accessories, gadgets and integrated appliances that it is easy to forget that they are still essentially workrooms, a place where food is to be safely and hygienically prepared. Of all surfaces in the kitchen the floor takes the brunt of hard wear. It must withstand grease, water, heat, heavy appliances and the concentration of foot traffic repeatedly standing in one spot or traveling over the same route between appliances and storage areas. It must be resistant to stains, easy to clean, hard-wearing and non-slip for safety. Despite these stringent requirements, both traditional and more recently developed materials offer sufficient decorative scope to indulge imaginative design ideas to the full, even if the kitchen must also be used for coffee with friends, relaxing with a good book or as the home office.

Unless already in situ, the kitchen floor tends to take its lead from the style of the units, but bear in mind its special needs and the fact that size and layout of design could significantly affect the visual dimensions of the room. With units built in around the walls and a large expanse of floor space to fill, for example, a small, light-colored all-over pattern seems to create a desolate prairie of emptiness that can't be filled with a piece of furniture for fear of causing a dangerous obstruction. A busy design or large pattern in bright colors could completely dominate a small galley kitchen. Generally speaking, the problem is one of filling that floor space with something decorative, even in small kitchens, and checks, stripes, mosaics and geometric designs remain perennially popular as optical fillers.

Where the room is designed to look fully functional – sleek and efficient as a spaceship control room or aiming at ultra-modern anonymity with everything hidden behind fascia doors – smart, practical surfaces like ceramic floor tiles and vinyl sheet or tiles offer unlimited choice of design options, from pure white or dramatic black to strong patterns and subtle shade combinations that can be continued over furniture and up walls for the smartest

Previous page Highly polished wood gives unexpected warmth and texture in this otherwise sleek and shiny hi-tech kitchen.

Above Hexagonal terracotta tiles and a dramatic double arch give this kitchen an unmistakable rustic feel.

kind of co-ordination. Smooth shiny surfaces like these are perfect for modern minimalism, and in a kitchen are ideally practical for their easy-care properties.

Hi-tech, clean-machine styling, so well suited to the kitchen as serious workroom/laboratory, can be taken a step further with flooring by adapting industrial materials to domestic use. Rubber is becoming increasingly popular. Manufacturers have extended their range of colors from broody black to brilliant primary reds, blues, greens and yellows. Extremely tough and completely waterproof, rubber flooring with its pattern of round, square or triangular studs makes an excellent self-patterned, non-slip surface.

Those seeking a strong, one-color floor that needs little maintenance could opt for plain concrete, given a couple of coats of tough concrete paint to produce a jewel-like finish. It's cheap and cheerful and potentially very stylish, especially if you have the patience to devise your own patterns in different colors. Or what about a metal floor to develop the spaceship, galley kitchen theme? You can't get smarter or sleeker than a shiny silver floor in stainless steel or aluminum sheeting, especially

Above Small, shaped tiles would have looked too fussy in this magnificent large kitchen, but the generous size and warm variation of natural shades in the squares, combines an unmistakable farmhouse look with up-to-the-minute convenience.

with a hi-tech worktop to match. Essentially practical and maintenance-free, this kind of reflective surface could be ideal for small kitchens where there's only room for the absolute basics and every inch counts.

More often than not though the kitchen is also supposed to be a living room, hobby room or dining area. Where the room is truly dual-purpose, a more utilitarian all-over flooring can be softened with rugs around a table or sofa, for example. The alternative is to furnish the dining area with a softer, more luxurious type of flooring material complementary to your kitchen flooring, but

with the choice of good-looking, practical surfaces available, this may not be necessary. Perfect companions for old-fashioned farmhouse pine or oak, for example, are old stone flags or mellow, russet clay or terracotta tiles – bricks laid in patterns or concrete garden pavers as a cheap alternative. You can buy cleaned-up originals or faithful reproductions in a variety of sizes and shades and, to add a little tasteful decoration, mix colors and shapes or add contrasting inserts and patterns. This is the way to install a floor that looks as though it has been there for years. Originally the floor would have been laid with local stone probably continued right through the house in large slabs or smaller pieces arranged in patterns. Those who want a truly authentic floor for their old colonial home or farmhouse should research and seek out local sources before making the final decision.

AUTHENTIC PERIOD FLOORS

The real thing is the only authentic accompaniment to top-quality, solid wood, traditional kitchen furniture in a period setting. But stone, slate and clay tiles are cold and hard underfoot, they are expensive and heavy to lay, and you can wave goodbye to anything you drop on them as it will certainly smash or shatter. If it is only the farmhouse look you are aiming to achieve, imitating the waxed wood finish and decorative styling of the cabinets and dressers, select one of the better vinyl flooring reproductions, marketed in wide sheet or tile form. These provide a much softer, warmer surface with the same visual impact.

All kinds of mock effects are available for kitchens including antique-look terracotta tiles, stones and slates, as well as fake marble and ceramic designs. Some are strengthened and cushioned for extra comfort and a more three-dimensional look. It is always worth paying a little more for the better products: they appear more realistic and will last longer. One of the big drawbacks to vinyl flooring in kitchens is that it can easily be damaged by large appliances when they are moved across the floor. This damage can be prevented by mounting large refrigerator-freezers or washing machines on special rollers, if they are not already fitted with castors.

Vinyl also produces passable mock marble and ceramic tiles. These again are softer and easier to handle, less expensive than the real thing and come in a wide choice of popular styles to suit, for example, a sunny Mediterranean atmosphere with fresh blue and white tiled effects; cherry red, or black and white 1950s checks or nineteenth-century marbled elegance with tiny black inserts.

If you can afford it and you like some texture underfoot, genuine ceramic tiles offer a wide scope for design possibilities. Plains and patterns can be mixed and matched into borders and panels; there are now many reproductions of Victorian picture tiles; look for hand-

made and painted Spanish and Mexican tiles – uneven and irregular to produce an instant mellow look; satin-smooth modern squares, bars and hexagons, or tiles ranging from 12in square right down to tiny mosaics. This is the kind of flexibility the synthetic options have yet to equal, and for anyone keen to try out creative ideas, ceramics are unrivalled for durability.

THE SOFT OPTION

Stone and ceramic, even soft shiny vinyl, can still be too cold and clinical for some tastes. These are the people who will often opt instead for cork tiles in the kitchen – not renowned for being real stunners but providing an excellent foil for country-style wood and available in plain white as well as bright primaries. They are also highly practical once sealed, easy to lay and comfortable underfoot. Cork's natural color is pleasantly mellow, but it is also available in other close shades, including gray – beautifully neutral against pastel or hand-dragged kitchen units. For more sophisticated tastes, a new wood floor in pale beech or limed oak is a beautiful companion for handpainted or softly colored units, or create a more traditional atmosphere with mellow polished pine or oak boards.

There are special carpet ranges designed to take the wear and tear of kitchen use. Styles are improving but, generally, the choice is mostly plain colors for all-over, wall-to-wall coverage. Such special carpeting is worth considering where units and other furnishings are dominant and you want a plain background with a little extra comfort. Carpet tiles are more interesting: you can mix and match colors and designs, make borders, checks or fit matching tiles invisibly together.

Soft and comfortable underfoot, easy to lay and quick to clean, vinyl also offers unlimited scope for design. A muted all-over pattern can be laid wall to wall to balance a strong choice of kitchen units (right), or cut and positioned to create exciting designs like the interesting grid arrangement of imitation wood and small ceramic tiles (left).

Left Kitchens as clean and bare as the inside of a space capsule cry out for a ceramic tiled floor. The hard, glossy or satin-finish surface suits the style perfectly. Plain pastel shades or hi-tech colors can suggest a futuristic feel and be smoothly co-ordinated with other surfaces such as worktops and walls.

Inset right Floor to ceiling wooden slats need strong companions to prevent them being overpowering. The solution is a dramatic midnight blue floor of ceramic tiles brightened by a brilliant red work surface to add both spice and drama to the arrangement.

Right Italian-style simplicity of line and color may lend itself to marble floors in the kitchen, but where budgets and safety prevail, faking the effect in vinyl can be equally effective and a lot more manageable. The soft, natural effect of marble is more interesting than plain white or gray and looks instantly luxurious.

Below Vinyl tiles imitating stone or marble can lift a plain treatment into the realms of something more classic for very little cost and effort. The continuous run of work surface and units in this kitchen naturally puts the floor in a position of some importance, and it required an eye-catching treatment to offset the white paint and the blank stretch of sealed wood.

Right Smart wood units with a sleek wood floor prove that natural wood needn't look folksy in the kitchen. Wooden planks or strips are comfortable and warm underfoot, yet they can be cleaned easily in busy work areas. This floor looks good enough to continue through into an informal dining area.

Above The minimal kitchen hides functional features behind glossy white cabinet fronts and covers the floor to match in large white vinyl floor tiles. It is smart, easy to clean and can be given a new look depending on the choice of colorful accessories.

Natural materials such as stone, slate, cork and wood have a subtlety of tone and texture that can be useful in the kitchen to offset the harder, shinier surfaces of appliances and laminated units (far left and below). Alternatively, you can play up the glossy, efficient look by using ceramic tiles in plain colors and large simple squares to emphasize their elegance (right and near left). The direction in which you lay the tiles is vitally important to the final effect, visually altering the dimensions of the whole area.

Left A sleek, modern kitchen takes advantage of the latest easy-care materials. A stylish black and white treatment has been continued on the floor with designer vinyl in imitation marble. The material has been cut and laid to produce the familiar cut-corner design, with a narrow double border of black.

Below French country style cries out for old-fashioned units and a natural tile floor whose honeyed warmth and variation of shades blends perfectly with the grain of the wood.

Left Woodblock can be laid to create an infinite variety of patterns. In the kitchen-diner, a centralized design draws attention to the table and chairs. The floor is decorative and practical in a dual-purpose room.

Left Warm and soft under foot, sealed cork tiles are a sensible choice in the kitchen. You can buy cork tiles in a variety of shades and effects: these are ready-coated in vinyl for ease of care.

Top right
Easy to clean but stylish, vinyl looks good and performs well. Plain or patterned, large areas can be quickly and comfortably covered. A wide choice of styles makes it possible to co-ordinate the floor with other decorations.

Far right Continuing the same floor surface through from the hall to the tiny galley kitchen helps increase the impression of space. It needed to be a tough, interesting and attractive floor to cope with the varying needs of passageway and work area. Woodblock imitation in vinyl fitted the bill perfectly and adds a warm glow to an otherwise glossy, monochromatic treatment.

Right Wooden floor and kitchen units have been subtly stained in a warm, light tone of cream. The theme is continued into the hall where the woodwork is stained with gray.

For an old-fashioned look in the kitchen, choose old-fashioned materials for the floor. If you are not lucky enough to have the original stone slabs, worn by years of use but still in excellent condition, like Fairfax House in England (left), you could always install genuine French provincial tiles (below). These are from a commercial range that were rescued from a French farmhouse and are between 100 and 200 years old. Providing tiles are not laid in cement, they can be lifted undamaged and cleaned ready for re-use.

Below Tiles can vary in color and texture from straw, ocher and brown to pinks and russets, creating a wonderful mellow patchwork.

Right An original floor is worth keeping even when slightly damaged, if only to maintain an authentic atmosphere.

Left Terracotta tiles can suit a multi-purpose room. Their natural earthy colors fit into both country-style kitchen and conservatory dining room. They are tough enough to take the wear and tear of a cooking area and garden room.

Below Judicious use of wood has breathed life into a plain white and cream kitchen. Polished wood has been chosen for both worktop and floor to give a sense of unity, reinforced by a wall cupboard and simple dining furniture. The wood softens the minimalist effect and makes the room more homey.

Right A Mediterranean-style walk-through kitchen is unusual, yet still practical, with the kitchen units arranged on either side and an eye-catching mosaic floor to provide an authentic atmosphere. In fact the floor is a fake, a top-quality vinyl imitation.

Chapter Seven

B E D R O O M S

A PERSONAL STYLE

Bedrooms are private, personal places and can be decorated accordingly. So far as the flooring is concerned, wear and tear is usually minimal so you can afford to be extravagant with thick-pile carpets or rare rugs.

The convention of running the stair and landing carpet straight through into the bedrooms isn't very exciting and, even with a neutral choice, limits bedrooms to similar design schemes. The carpet can be used as a background to different rugs to ring the changes, but much more fun is to plan something special for each room according to the atmosphere you wish to create. Somehow the bedroom lends itself to flights of fancy without being impractical or looking silly and you don't have to worry about the floor covering being super tough or stain-resistant so long as it's warm and comfortable underfoot.

Sometimes the style of the house or a particularly grand or decorative bed will set the decorative scheme and suggest a floor material. The fabulous four-poster bed in country cottage or grand manor house looks good with polished boards and oriental rugs casually strewn to echo the colors of a patchwork quilt. Traditionally the carpet ran in strips around the bed, so at least your feet didn't touch down on a cold floor first thing in the morning. Add more rugs, attach a few to the wall, pile the bed with cushions, and turn down the lighting to create an exotic atmosphere. Install a pure white, long-pile wall-to-wall carpet and a glamorous headboard and you have instant Hollywood opulence.

Bedroom styles can be the last word in luxury with a thick, soft carpet. Others manage to reproduce the starkness of Provençal rooms or simple Shaker bedrooms with stripped and polished or bleached boards, a plain natural wood or brass bed and rough white or color-washed walls. Decoration here will be restricted to stenciled borders and designs painted directly onto the boards and simple grass mats or cotton rugs. A home-made floor cloth would also be in keeping and add originality.

A fantasy bedroom with elaborately paved, marbled

Previous page Traditionally bedrooms would be furnished with exotic rugs for warmth and comfort on stone or wood floors.

Right A sumptuous setting with elaborate drapes and splendid fabrics has a floor covering that doesn't compete with other furnishings – wall-to-wall carpet in soft cream to enhance limed chestnut and honey-colored decorations.

Below The soft pastel bedroom with its pale painted cupboards and low-key furnishings, has a large dhurry in perfectly matching pastel shades to add interest to the plain carpeted floor.

or tiled floors wouldn't be practical, but with the aid of modern vinyl reproductions, you might achieve the illusion simply by adding the right accessories. Any kind of hard surface can be softened with mats and rugs, positioned where they will be best appreciated by bare feet. All styles look good in the bedroom, especially the flatter, woven types like kelims and dhurries, which might not take the harder wear elsewhere in the house. Keep a look-out for rugs that seem to capture an intended atmosphere exactly – it might be tartans for a Scottish theme; genuine Japanese bamboo or grass mats for a simple oriental room complete with futon bed; or Chinese sculpted flowers in wool or silk to complement a delicate floral theme. Rugs are the perfect bedroom accessory!

CHILD'S PLAY

In children's bedrooms, the requirements are completely different. The flooring must be exciting and inspired, but luxury must be replaced by practicality in a room that often has to withstand the rough and tumble of doubling as a playroom. While areas of carpet are warm and comfortable for sitting at play for extended periods and provide a soft landing for occasional falls, carpets have to be tough and stain resistant when children are still young. Wall-to-wall carpet isn't welcome if small occupants want to explore the possibilities and limitations of building blocks, battery-operated cars, puzzles and other toys requiring a sound, stable surface. If you are worried about cold feet and comfort, carpet tiles might be a solution since they can be lifted individually for cleaning or to clear an area for games as required.

Bare boards will suffice, providing they are well sanded and checked for snags and nails. You could paint or stencil them with animal shapes and finish with plenty of coats of varnish. Another way would be to paint the boards alternate primary colors in gloss, or measure out an ever-decreasing border towards the center of the room and fill it in with bright colors. Boards are noisy, so a couple of rugs – buy them with favorite cartoon characters or alphabet figures on them – will help to reduce any disturbance. A false floor with sound insulation between is the perfect solution but is expensive.

Vinyl flooring is good for children's rooms. It is soft and warm, fine insulation and easy to lay in continuous strip or tile format. A small or an all-over neutral pattern, plain colors, checkerboards or stripes offer maximum play possibilities. The thick cushioned type of vinyl flooring is more comfortable but unlikely to receive a vote of confidence since it is too uneven to run small cars over or make the base for painting or other games. Beware too of bright, busy patterns – small component parts can easily disappear, to be trodden on later or sucked up in the vacuum cleaner. Equally warm and comfortable, and probably a better choice for a neutral, level floor that will see a toddler through to the sophistication of teen years, is cork tiles. They are tough and hardwearing but their natural mellow color is instantly appealing. Tiles can be stenciled with colored paints if you want to add something of interest, and the surface

Above A magnificent chestnut four poster bed and an antique polished wood floor are matched with a richly patterned rug.

should be simply resealed with varnish when it starts showing signs of wear. In bedroom or playroom, a cork floor continued up the wall to skirting level will protect the wall from tricycle and baby walker damage; a few spare tiles fastened to the wall make a good corkboard, too.

Hardboard is often used to create a level subfloor and provide extra insulation beneath tiles, vinyl and linoleum. If you lay it shiny side up and make sure the sheets are neatly butted together, the surface is easily painted and sealed with a pattern or design. This could take the form of an extensive road layout for toy cars, a miniature football pitch, a maze or board games such as snakes and ladders or checkers. Even an assortment of simple shapes and patterns in bright colors will stimulate young imaginations.

Left Vinyl flooring reproduces the effect of real marble in the bedroom, yet provides all the comfort of a soft, warm surface underfoot.

Above As the bedroom doesn't receive the wear and tear of other rooms in the house, you can enjoy the wall-to-wall luxury of a thicker, more expensive pile and plain, light colors.

Right A strongly patterned carpet can work in the bedroom if color and pattern are chosen carefully to co-ordinate with other furnishings. Since the general style of this bedroom is buttons, bows and frills, the pink and gray carpet is a perfect companion to the pink roses.

To recreate the elegance of a former age, the flooring needs to be chosen particularly carefully or it will ruin the effect of fine furnishings. Usually floors were left plain polished or scrubbed wood with exotic rugs for comfort according to money and position. The beauty of this arrangement (near left and right) is that rugs can be changed or replaced according to whim or finance. They certainly create a grand atmosphere in the bedroom.

Larger carpets can work well in the traditional bedroom providing they feature an appropriate design like the simple fleur-de-lis on the wall-to-wall carpet (far left), which picks up the design of the wallpaper. The exotic silk rug (below left) virtually fills the room, making an ideal setting for a dramatic bed and rich fabrics.

Right Rugs are a great idea for ringing the changes in children's rooms, adding bright colors and a soft surface to play on, yet they can be easily rolled up and put to one side if a hard surface is needed for games.

Above Older children might appreciate a stylish rug to soften a practical wood floor in a work/ playroom.

Right Soft and comfortable, yet easily cleaned and good for play, sheet vinyl makes the best all-over surface for young children. A plain or simple geometric pattern offers possibilities for games.

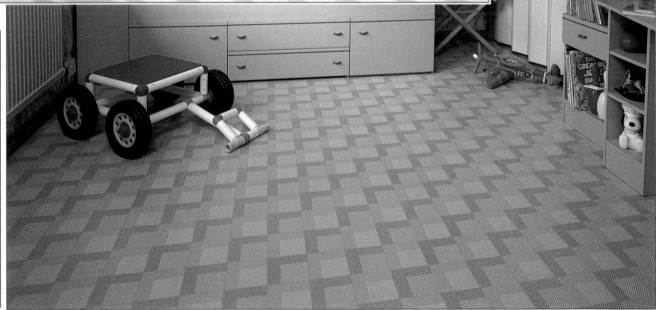

Left A plain, wall-to-wall flooring in the bedroom can set the mood or reinforce a theme: dress it up or down according to taste and whim. Rough-textured coir is perfect for a country-cottage bedroom with a matching blind and simple sprigged wallpaper. White-painted bed and plain white bedlinen maintain a feeling of freshness and austerity.

Left below For a cozier, cottage atmosphere, carpet was chosen. Texture and color are introduced with plain pine furniture and rich fabrics.

Right In the bedroom you can get away with the self indulgence of a large expanse of velvet cut carpet, almost like suede, to create a room of oriental minimalism.

Chapter Eight

B A T H R O O M S

◇

MAKING A SPLASH

The total floor area in the bathroom, once the fixtures are installed, tends to be so small that it is frequently neglected and simply covered as an afterthought. Yet its smallness is the perfect opportunity to experiment with more expensive materials or to be a little more imaginative than in the other, larger rooms, where to replace the complete floor would be costly. If the space to cover is very tiny, you can often take advantage of discounted remainders on ranges or offcuts, or spend the time devising and fitting your own designs, maybe hand-painting tiles or wood. The idea of handcutting and fitting linoleum or ceramic tiles, or buying marble or terrazzo is normally so alarming that the project wouldn't even be considered in the majority of homes. But in the average bathroom, reduction of scale makes most dreams come true.

There are limitations of course. Like the kitchen, the bathroom imposes strict requirements on all its decorative materials: they must fit neatly, be safe and comfortable underfoot and completely resistant to water splashes, steam, talcum powder and oils. If you have set your heart on some of the heavier flooring materials such as ceramics or marble and the room is above the ground floor you will have to check that the floor joists can take the weight of your chosen flooring and reinforce where necessary.

That isn't to say there is a lack of choice. Within the boundaries of ceramic tiles, cork, rubber, vinyl, even carpet and wood if it is specially prepared for the purpose, there's more than sufficient choice to create the desired look. Actually selecting the correct color, pattern and texture can be more tricky, small area or no. Where the walls are already tiled, and with the dominant natural glossiness of fixtures, a shiny surfaced floor could make the room cold and clinical-looking. The temptation to use matching ceramics over wall and windowsill, up the side of the bath and across the floor could be too much and reduce the room to having all the appeal of a public restroom. It can work, but it needs a keen eye for color and pattern to balance the effect and

a careful choice of softening accessories. It works better in a small room (and would be very expensive in a large one) where an all-over matching material can increase the size of the room visually. Large mirrors will also increase the possibilities of playing optical tricks. You should take care to choose from the plain and more subtly patterned styles – most ranges include co-ordinated, strong designs which can be used judiciously to make panels, borders and stripes to add highlights and points of interest. A band of pure color on white or similar background shade, continued down the wall and along the floor in matching ceramics, is highly effective. Leading tile manufacturers are now producing field (plain) and patterned tiles to match popular fixture colors, allowing all kinds of co-ordinated effects. Don't assume the bathroom has to be decorated in pastels; dramatic dark colors are often more effective and interesting, even for small rooms.

If the totally tiled room isn't to your taste or budget but you prefer a ceramic tiled floor, the answer may be to go easy on the wall tiles. Restrict them to splashbacks behind bathtub and sink or only tile the area above or below dado rail height (about waist level) – a wooden rail or narrow edging tile insures a neat finish here – or use tiles in selected areas to pull a design together: carry them up the side of the bath and use them to cover integral bench seating or shelving. In a large room, where a ceramic floor might be too dominant and chilly looking, or simply too expensive, tiles might still be incorporated where they would be most practical. Mixed with other softer, flooring materials – bathroom carpet in a complementary color is often the best choice – tiles could be restricted to the area around the bathtub (especially if it is sunk in or slightly raised), or within a step-in shower area. This technique also works well for master bathrooms

where the carpet runs through from the bedroom but stops short of any high-splash areas.

The approach must be the same when considering marble, mosaic or terrazzo. Limiting the area reduces the cost and effort: as an inset or section they can look extremely effective when mixed with the right materials. They can be slippery and are not recommended for bathrooms used by the elderly or very young. They can, however, be perfect for introducing a luxurious atmosphere to a master bathroom.

Trying to combine a sense of comfort and luxury with practical considerations affects choice and use of carpet in the bathroom too. It is important to use a type specially recommended for the purpose. These do exist in a good choice of colors and with a reasonably long pile that feels delicious under bare feet. Bathroom carpet can also be found to match standard carpet ranges so that it can be

Above Color co-ordination achieves a feeling of space in a tiny room. The sheet vinyl design used for the floor gives wall-to-wall neatness.

matched to the bedroom or landing floors if required. It is designed to withstand steam and splashes but it is not advisable to get it thoroughly soaked, as the fiber may still rot if it isn't dried out quickly enough. With young children it may be better to use the inset tile or vinyl idea around the bathtub and sink to reduce the amount of water reaching the carpet. Bathroom carpet is also available in tile form, allowing you to lift and dry specific areas should they become saturated.

For many though, the thick, soft texture of carpet doesn't sit happily with sleek fixtures and other bathroom fittings, or they prefer a smoother, more streamlined look and a more durable material. If ceramic or marble feels too cold, or is too heavy for the floor, there are always the vinyl designs allowing you the chance to recreate the effect of a Roman bathhouse, Mediterranean villa or whatever design takes your fancy –

indulge your imagination. Vinyl is easily fitted and feels soft and comfortable, withstanding splashes well. However, it needs to be well secured around the edges so that water cannot seep beneath it, and it is not suitable for shower cubicles or areas receiving a direct flow of water.

Cork tiles can be used in the same way as vinyl and they produce a pleasant mellow warmth that makes a good foil for white, cream, black, dark green or burgundy fixtures. It is a good choice for family bathrooms, being soft yet strong and practical: sealing makes it waterproof, but leaving it natural will absorb moisture

Below Edwardian opulence is provided by brass and mahogany and a classic marble-tiled floor. Marbles and ceramics can achieve superb effects on bathroom walls and floors.

from the air and reduce condensation problems. The beauty of cork is that it is adaptable and easily fitted to the side of the bathtub, to seats or any surface you feel it suits. Its natural color and finish tends to add a traditional air to the bathroom.

For a tough, up-to-the-minute floor in strong colors you could always use rubber stud flooring. This is another material that can look effective carried a little way up the wall or along the side of the bathtub. It comes in bright, dramatic colors for pairing with plain white or black bathroom suites, and the studded surface is not

unpleasant underfoot – it has a kind of massaging effect like Japanese health sandals. Other fun surfaces for bathroom floors include fake turf effects. Even more practical is duckboarding, the kind once used to protect floors at swimming pools. These raised slatted platforms are comfortable and practical; the water drains onto a waterproof surface beneath, leaving the top dry. Use it in sections close by the bathtub or shower. It may be hard on feet if it covers the whole floor.

Bare boards may be a tempting alternative for the bathroom, especially combined with one of those large,

free-standing roll-top baths and a scattering of rugs to create an old-fashioned or minimalist atmosphere. Boards can be stained or stenciled and well varnished for protection – an inexpensive flooring choice for large rooms. But remember that the boards must be sanded scrupulously smooth if there is to be no risk of splinters and that water spills may seep through and ruin the ceiling below.

THE BATHTUB MAT

The duckboard can be a practical alternative to the bathtub mat, and there are other stylish options instead of the ubiquitous rectangle of towel matting. This is not the place for beautiful Persian rugs, but an inexpensive oriental-style rug, or a lighter dhurry or kelim which can be hung to dry quickly, looks splendidly luxurious in a traditional-style bathroom with a tiled, natural wood or cork floor. Also offering a touch of class are the self-patterned white cotton rugs which are easily put in the washing machine or tumble drier. With all mats in the bathroom, you should insure they are firmly anchored and unlikely to slip.

Below right Often the most successful treatments in bathrooms are designed with walls and floors as one, like this moody marbled room.

Left Plain tiles are less expensive than patterned ones. Create your own patterns with tiles in two colors.

Left Monochrome simplicity is made possible by continuing plain floor tiles up the side of the bathtub and along walls and into a walk-in shower. This treatment makes a small bathroom look bigger. An all-white treatment , matching tiles to fixtures and painted surfaces, is a relatively easy task, but the size of the tiles is crucial if the effect is not to look too institutional.

Right Surprisingly, a small room can take strong treatment with floor and walls tiled in deep brick red, relieved only by the judicious addition of white. By adding a third shade and taking the colors of a simple rug as inspiration (far right), the bathroom takes on a softer, no less impressive look, with tiles providing a neutral background for the rug itself, but echoing its color around the walls.

Below right Visual illusions with all-white tiles and clever use of mirrors are created in this bathroom, which has plain flooring tiles running around every available surface including windowsills and the sunken bathtub. With green plants and a touch of shiny chrome the only accessories, this design relies on different shaped tiles for variety and interest. Small interlocking pavers on the floor give a completely different effect to the conventional square-cut wall tiles.

Left Ceramic tiles come in so many wonderful colors and patterns, you can reproduce any effect in the bathroom, secure in the knowledge that it will withstand heat, water and soap suds. The elaborate pattern of small tiles picks up the strong contrasts of heavily stained and polished wood against a ceramic tiled wall and heavy marble and white fixtures. All conspire to create a moody Edwardian atmosphere that would have been ruined by an insipid flooring.

Left A star-burst pattern effect in large tiles creates a soft floor design in muted shades. In earthy tones of sand and terracotta, the floor becomes the focal point of the room, helping to balance a tall ceiling and warm up the effect of stark white fixtures and cool blue paint.

Right Not a Persian carpet, but tough and practical sheet vinyl in an exotic bathroom setting – this bathroom floor was the starting point for this treatment, suggesting the shade for painted walls and old-fashioned free-standing bathtub. Even the towels have been chosen to match the deep blue of the design. A wealth of extravagant accessories and plenty of glossy green plants reinforce the atmosphere.

Left Since the floor is bound to be the dominant feature in the bathroom, play it up or down according to the impact of other features in the room. Thus this moody room, with its deep black tiles and orange marbled wallpaper with woodwork, radiator and paintwork to match, required the soft neutral swirl of a cushioned vinyl.

Top right Contrasting tiles can always be relied upon for a note of drama and exciting floors full of geometric interest. Black and white vinyl is the perfect companion for 1930s-style fixtures and accessories, whether they are cut corners with black inserts (near picture) or full-sized black and white tiles.

Right Ceramic tiles are not only tough and easy to clean, they create the most authentic antique atmosphere in an old-fashioned interior. This Victorian-style design in earthy colors is a classic and makes a stunning background to plain white fixtures.

Left Ceramic floor tiles will make the most of small spaces. These specially laid hand-painted examples match up to the ornate fixtures.

Above Ceramic floor tiles have been cut and fitted into a simple geometric design.

Right This incredible gothic interior has a stone floor and exquisite Persian carpet. It can certainly be described as an elegant and sumptuous retreat.

Chapter Nine

CONSERVATORIES AND OUTDOORS

◇

INSIDE OUT

ntrances and exits can so easily fall miserably between two stools and satisfy neither the practical requirements of outdoors, nor match the stylishness of the interior. It may be a tiny porch condemned to nothing more than its original concrete subfloor through lack of interest, or a conservatory/sunroom at the rear, furnished with off-cut vinyl or indifferent paving.

They had the right idea in the nineteenth century when even the humblest, smallest entrance porch would be proudly decorated with a patterned floor in stone, slate or ceramic to match the path leading to the door. Much of this handiwork is still about, surprisingly intact considering the wear and tear it has taken – checks, diamonds and borders in earthy reds, browns, yellows and black. This is the sort of effect to aim for in general when you decide on a floor – a harmonious transition from indoors to outdoors that looks pleasing and will visually extend the living area into the yard or garden.

Of course a floor design should not be restricted to Victorian-style geometrics. Providing the chosen material is tough, waterproof and weatherproof – especially if you are planning to continue it into the yard or onto the patio – the only limitations are budget and imagination. Naturally materials designed for paths and patios are ideal and the choice here is surprisingly wide and stylish – stone, slate, brick and natural wood in all manner of shapes and colors. There is certainly plenty of scope even if you decide on a less weather-resistant flooring such as ceramic tiles or terracotta inside, and devise a complementary pattern and texture outside using something more suitable. You could extend the idea of a border, for example, or a pattern of tiles could become a design in brick. It is the perfect opportunity to play around with interesting and contrasting textures – the smooth shininess of terrazzo or ceramics, the roughness of brick or stone, the difference between polished and rough-sawn wood. Providing pattern and tone are kept within the same style, you can achieve a pleasing harmony with contrasts such as these.

It pays to think through the practical limitations of your area before you install any kind of flooring. A porch may need to house wet umbrellas and muddy shoes, roller skates and a pet flap, wet bathing suits and tennis rackets. Incorporate a sunken well for a doormat if you can or lay a wall-to-wall mat at the door.

Conservatories are popular for use as an extra dining room and with this aim uppermost, the floor is frequently unwisely chosen. Remember that a conservatory is not a true conservatory without plants and that these will need regular watering. To maintain the correct humidity levels in summer, it also helps to hose down the floor daily, and this requires a highly waterproof surface and integral drains. If you want to soften such a floor, use easily cleaned rugs.

Once you have identified how the area will be used, you can start exploring the creative options. As an introduction to or extension of the home, it makes sense to go for something that reflects your general taste and lifestyle, no doubt in evidence around the rest of the house.

Traditionalists will be looking at mellow terracotta tiles or ceramic mosaics, maybe marble with smart black inserts. Both original antiques and quality reproductions are available, from the country farmhouse look to the elegant townhouse. Those seeking eye-catching modern patterns or Mediterranean coolness will want to experiment with ceramic floor tiles in large and small sizes, with borders, strong patterns and subtle pastels in every color and shade imaginable.

It may be the yard that sets the style, with stone slabs,

bricks or pavers continued from outside. An increasingly popular alternative here is wood decking; softer underfoot and easier to install, it can be laid across any area, into any shape and at any height. It looks most effective arranged in patterns such as herringbones and weaves, varnished to bring out the natural color of the wood (both hard and softwoods can be used), or stained a bleached-out white or gray, strong red, blue or green according to taste. Continued into a porch or enclosed terrace area, the wood might be stenciled or patterned with colored stains for a more decorative effect.

DESIGN BRIEFS

Porches are often small, offering the chance to go to town with something expensive and create an impressive introduction to the house. A small design such as mosaic or terrazzo is perfect for minimal areas: imitate the popular Victorian florals or design your own pictures, maybe something witty like the famous snarling dog and mosaic inscription *CAVE CANEM* (beware of the dog) at the entrance to one of the villas at Pompeii. The usual square or rectangular shape of the average porch also lends itself to terracotta and ceramic tile designs, with or without the classic clipped corners and colored inserts of marble or clay. Alternatively, set the design on the diagonal or use stripes to disguise the true shape and size of the area. Brick, available in reds, yellows and black, is also useful for hard-wearing decorative designs like basketweave and herringbone. Avoid carpet. Any area acting as buffer between the house and the outdoor world takes extremely hard wear; it acts as a filter for most of the dirt and dust walked in on feet and paws. If you really don't want a surface that has to be swept and swabbed, a possible alternative is coir matting. It's rough and tough but still decorative, particularly some of the new weaves and patterned colors. The dirt tends to drop through so the matting remains looking good, but will need to be lifted regularly for cleaning beneath.

Conservatories offer a much larger area to play around with and present a totally different design requirement. As mentioned earlier, the floor needs to be practical to cope with the rigors of plant care, but that soaring space full of glass and light and the absence of any decorative distractions such as fabric or wallpaper (even the furniture

must be restricted to bamboo, rattan or outdoor types – no highly polished woods that are not water-resistant), brings the floor into focus.

Tiles or stone slabs are probably the ideal – the perfect compromise for the needs of both people and plants. The nineteenth-century leads the field in stunning design here: checkered marble, mosaics, borders, pictures and panels, all can be achieved with marbles and ceramics. Since the majority of conservatories are based on original nineteenth-century designs, a similar style is bound to look good. An all-over pattern is probably most appropriate for small structures, with maybe a border for added interest. But the larger, usually circular or hexagonal conservatory should take its lead from the roof. Whether square, rectangular or circular, centralize your design below the main roof structure and fan out towards the sides. At night, lit by lamps or candles, the floor will be exactly reproduced in the roof glass – a truly stunning effect for dinner parties when table, crystal and silverware will be similarly reflected.

With the variety of designs and colors available, a ceramic tile or terrazzo design can be as understated or spectacularly decorative as you wish. For those who prefer a more subtle, country atmosphere, less gloss and a closer affinity to the plants that contribute the only other main decoration in a well-planned conservatory, choose earthy terracotta tiles, slate or stone flags. They come in large sizes so big areas can be quickly and effectively covered, avoiding the more broken effect of mosaic or smaller ceramic tiles. You can mix sizes and shades to create subtle pattern variations or add inserts of a different material or color. This is often the best option for an enclosed terrace room where the narrowness of the building makes too bright and busy a design overpowering. After all, you are supposed to be concentrating your attention on the view.

Here, and in the sunroom – more of a room than a plant house like the conservatory – plants tend to be restricted to containers rather than direct bed planting or staging. Elegant Versailles tubs, of stout terracotta or sculpted stone, could be stood successfully on a wood floor, maybe wooden decking continued through French windows or large glass doors onto a patio area. Oriental rugs, grass mats or woven kelims might be used to soften and add a touch of comfort to any of these floors.

Left The mosaic effect of tiny stones in two shades has made a perfect inside/outside flooring surface. Pure Mediterranean in feel, cobbles require plenty of patience to design and lay, but the finished effect is beautiful and highly practical.

travels well around awkward shapes, across water or between planting beds, into the house or out onto the patio. This exotic and ambitious conservatory (below) has used wood magnificently: matching wood is used for floor decking laid on the diagonal; for a stout enclosure and handrail following the route of decking and seating areas; for dining furniture; and for the construction of the conservatory itself. The wood has a natural warmth and softness that is perfectly balanced by a profusion of cool greenery. A similar technique has been employed (right) with complete wood treatment on floors and ceilings, walls and furniture. Again the wood has been carried straight through to the decked area outdoors, the indoor area being given a strong color and finish to create a contrasting decorative effect.

Above The comfortable conservatory sunroom puts the emphasis firmly on cozy wicker furniture and a fine view. A large dhurry is ideal to add warmth and color to the terracotta-tiled floor, and it can be easily lifted for cleaning or sweeping. Its flat weave is also better suited to this indoor/outdoor environment than a thick pile carpet or rug which would quickly accumulate dust and dirt and then need expensive cleaning.

Above Slate and brick blends well with wood – a rusty hue and a blue gray – as in this unusual garden room. The floor needed to be strong to balance the bold wood and brick construction. The added advantage is that it is tough, withstands water and copes well with plant debris and other stray material.

With conservatories, French windows and patio doors making the transition between home and garden or patio less distinct, outside paving and flooring materials can be as creative as those used inside. Strike up a contrast with rough and ready slabs (left); continue a path right up to the door and beyond (far right); or continue your indoor surface outdoors like the stylish ceramic-tiled roof terrace (near right), with neatly clipped plants in containers and rugs and cushions for comfort.

Any outdoor material such as stone or brick is perfect for porches, conservatories, entrance halls and passages, since it can be continued right into the house from a path or patio. It looks stylish and both areas appear bigger than they really are. Indoors, paving slabs, stones and setts can be sealed with flooring sealant and polished to a fine finish. But don't let your imagination stop at the porch or garden door: create an informal small country-style backyard with irregular flags and soften their edges with plants (left). Or experiment with patterns and shapes using small pavers, brick or tile (right). Use foliage and ground-cover plants as you would rugs and other furnishing accessories to soften and to add color and interest.

Terracotta, stone and slate make the perfect conservatory floor: they are smart enough to allow you to dress the room up a little, yet will withstand the mess and moisture of being situated in a plant house. Although the natural coloring of fired clay or quarried stone is subtle, it offers plenty of exciting design possibilities; you can mix shapes and shades or add colored inserts. Cover large areas with new but already mellow terracotta tiles (left); or install antique tiles for an old-fashioned atmosphere like the superb French hexagons (below), or the multi-shaded tiles (right).

INDEX

ACKNOWLEDGMENTS

◇

Quarto would like to thank the following for their help with this publication and for permissions to reproduce copyright material.

Key: A—Above; B—Below; R—Right; L—Left; C—Center; T—Top; Bt—Bottom

Every effort has been made to trace and acknowledge all copyright holders. Quarto would like to apologize if any omissions have been made.

AFIA Carpets 64 (L), 71, 82 (A), 125
Amtico 35, 36 (BL), 37 (L), 119
Arcaid 44, 50 (B), 74, 82 (B), 130 (B)
B.C. Sanitan 7 (B+L), 136, 143 (AL, AR), 145
Berleigh Publishing 33 (B), 40 (R), 46 (R), 50 (AC), 68, 69 (B), 94 (R), 104, 116, 117 (A) 126 (AL, AR), 145 (L)
Trevor Caley 33 (AL, AR)
C.M. Dixon 10
Fired Earth 116 (B)
Forbo-Nairn 36 (AR), 128 (B)
Michael Freeman 18, 19
Hillfield Studios 144
Hutchinson Library 9, 25, 28L
Ibstock Building Products Ltd 7 (BR), 43 (A)
Ikea 51 (B)
H+R Johnson 41 (A)

Junckers 45
Michael Kennedy 48, 50 (AL, C)
G. Kievel + Sons 26
Leeds Castle Foundation 11
Leighton House 14-15
Marley Floors 105, 135
John Lewis of Hungerford 103
National Magazine Co, Good Housekeeping; Tom Leighton 34 (B), David Brittain 49, 153 (L), Jan Baldwin 77 (B), Peter Anderson 90, 134, 142, Dennis Stone 112
Paris Ceramics 6 (AR), 27, 37 (R), 38 (R), 41 (AR), 42, 55, 57, 113 (B), 118 (A), 156 (B), 157
Pipe Dreams 86
Smallbone 94 (L), 95, 102, 117 (B), 122 (A, B), 123, 126 (B)
Barrie Smith 64 (R)

Sussex Terracotta 38 (L), 39, 40 (BL), 41 (B)
Spanish Tourist Board 23
Tomkinsons Carpets Ltd 50 (AR)
Town + Country Conservatories 2, 3, 87 (B), 147, 148, 156 (A)
Trio Design 54 (L), 87 (A)
Victoria + Albert Museum 22
Elizabeth Whiting Associates 5, 6 (BR, BL), 7 (CL, AL, AR), 28 R, 29, 30, 31, 34 (A), 36 (AL), 40 (AL), 46 (R), 47 (AL, L, R), 51 (A), 53, 56, 58 (L, R), 59 (AL, AR, B), 60 (L, R) 61, 62 (L, R), 63, 65, 66 (L, R), 67, 69 (A), 72, 73, 75, 76, 77 (A), 78 (A, BL, BR), 79, 80 (A, B), 81 (A, BL, BR), 83, 84 (A, B), 85 (A, B), 87 (C), 89, 91 (A, B), 92, 93 (A, B), 96 (A, B) 97 (L, R), 98 (AL, AR, BL, BR), 99, 101, 106, 107 (A, C), 108 (L, R), 109, 110 (AL, AR, B), 111, 113 (A), 115 (TL, TR, B), 118 (B), 121, 124 (L, R), 127, 128 (A), 129, 130 (A), 131, 133, 137 (L, R), 138, 139 (AL, AR, B), 140 (A, B), 141, 143 (B), 149, 150 (AL, BL, R), 151, 152, 153 (R), 154, 155
Wicanders 46 (L), 114
York Handmade Brick Co 43 (B)